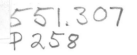

PARTNERSHIPS

FOR REDUCING LANDSLIDE RISK

Assessment of the National Landslide
Hazards Mitigation Strategy

Committee on the Review of the National Landslide Hazards Mitigation Strategy
Board on Earth Sciences and Resources
Division on Earth and Life Studies
NATIONAL RESEARCH COUNCIL
OF THE NATIONAL ACADEMIES

THE NATIONAL ACADEMIES PRESS
Washington, D.C.
www.nap.edu

THE NATIONAL ACADEMIES PRESS • 500 Fifth Street, N.W. • Washington, DC 20001

NOTICE: The project that is the subject of this report was approved by the Governing Board of the National Research Council, whose members are drawn from the councils of the National Academy of Sciences, the National Academy of Engineering, and the Institute of Medicine. The members of the committee responsible for the report were chosen for their special competences and with regard for appropriate balance.

The views and conclusions contained in this document are those of the authors and should not be interpreted as representing the opinions or policies of the U.S. Government. Mention of trade names or commercial products does not constitute their endorsement by the U.S. Government. Supported by the U.S. Geological Survey, Department of the Interior, under assistance Award No. 01HQAG0193.

International Standard Book Number (ISBN) 0-309-09140-3 (Book)
International Standard Book Number (ISBN) 0-309-52995-6 (PDF)

Additional copies of this report are available from the National Academies Press, 500 Fifth Street, N.W., Lockbox 285, Washington, DC 20055; (800) 624-6242 or (202) 334-3313 (in the Washington metropolitan area); Internet http://www.nap.edu

Cover: Image courtesy of David A. Feary; cover designed by Michele de la Menardiere.

THE NATIONAL ACADEMIES
Advisers to the Nation on Science, Engineering, and Medicine

The **National Academy of Sciences** is a private, nonprofit, self-perpetuating society of distinguished scholars engaged in scientific and engineering research, dedicated to the furtherance of science and technology and to their use for the general welfare. Upon the authority of the charter granted to it by the Congress in 1863, the Academy has a mandate that requires it to advise the federal government on scientific and technical matters. Dr. Bruce M. Alberts is president of the National Academy of Sciences.

The **National Academy of Engineering** was established in 1964, under the charter of the National Academy of Sciences, as a parallel organization of outstanding engineers. It is autonomous in its administration and in the selection of its members, sharing with the National Academy of Sciences the responsibility for advising the federal government. The National Academy of Engineering also sponsors engineering programs aimed at meeting national needs, encourages education and research, and recognizes the superior achievements of engineers. Dr. Wm. A. Wulf is president of the National Academy of Engineering.

The **Institute of Medicine** was established in 1970 by the National Academy of Sciences to secure the services of eminent members of appropriate professions in the examination of policy matters pertaining to the health of the public. The Institute acts under the responsibility given to the National Academy of Sciences by its congressional charter to be an adviser to the federal government and, upon its own initiative, to identify issues of medical care, research, and education. Dr. Harvey V. Fineberg is president of the Institute of Medicine.

The **National Research Council** was organized by the National Academy of Sciences in 1916 to associate the broad community of science and technology with the Academy's purposes of furthering knowledge and advising the federal government. Functioning in accordance with general policies determined by the Academy, the Council has become the principal operating agency of both the National Academy of Sciences and the National Academy of Engineering in providing services to the government, the public, and the scientific and engineering communities. The Council is administered jointly by both Academies and the Institute of Medicine. Dr. Bruce M. Alberts and Dr. Wm. A. Wulf are chair and vice chair, respectively, of the National Research Council.

www.national-academies.org

v

Preface

. . . the mountain falls and crumbles away . . .
and the rock is removed from its place

Job 14:18

Almost every part of the world is subject to landslides. Wherever there are mountains, or even hills, there have been, there are, and there will continue to be landslides. Landslides are a component of the erosion process, a continued leveling of the surface features of the earth—both on land and beneath the sea—that are thrust up by the collision of tectonic plates. Landscapes are shaped by such erosional processes, most dramatically by landslides.

Henry David Thoreau remarked, "The finest workers in stone are not copper or steel tools, but the gentle touches of air and water working at their leisure with a liberal allowance of time." However, there are times when the finest workers are not very gentle, when the natural sculptor seems to be angry and impatient, flinging large pieces from the emerging landscape, the sooner to finish the work. When people are in the way, the natural process is termed a catastrophe. Its imminence constitutes a natural hazard to mankind.

As the world's population has increased rapidly over the past century and as people move onto previously uninhabited land, there has been a greater interaction between humans and landslides, often to the detriment of humans. Just as people who live in earthquake zones, and who do not understand earthquakes, rebuild with the same erroneous methods of the past, so do people who challenge nature by building and living in harm's way in landslide zones without understanding that the control of nature is, in the long run, not possible.

Thus, this report is not about the prevention of landslides. What it is about is improved understanding of the hazards posed by landslides, of the role that improved education and the dissemination of information can play, and about the mitigation of such hazards through improved building and inspection codes and through improved engineering practice. The identification and assessment of landslide hazards and the evaluation of the risks associated with acts of mitigation are discussed in this report from two points of view. First is the objective point of view of the natural sciences, and second is the subjective point of view that people have to understand the bargain they make with nature when they choose to live in rugged terrain.

J. Freeman Gilbert
Chair

Acknowledgments

The committee would like to express its appreciation to the many individuals who provided briefings and other information during the information-gathering process: Karen Berry, Steve Briggs, Peter Bobrowsky, Scott Burns, Vicki Cowart, Jerome V. DeGraff, Christopher Doyle, Jerry Fish, Richard Fragaszy, Paula Gori, Edwin Harp, Rex Hickling, Jerry Higgins, Sanjay Jeer, Jeffrey Keaton, Pat Leahy, Paul Logan, Mike Long, Steve Olson, John Pallister, Donald Plotkin, Robert Schuster, Barry Siel, Lawson Smith (deceased), David Steensen, Joan Van Velsor, Yumei Wang, Jeffrey Weissel, Gerald Wieczorek, and William Ypsilantis. The committee particularly acknowledges the provision of information dealing with the International Consortium on Landslides and other international activities by Robert Schuster and also wishes to thank Yumei Wang, Scott Burns, and their colleagues for providing the committee with first-hand information on landslides in the Columbia Gorge area.

This report has been reviewed in draft form by individuals chosen for their diverse perspectives and technical expertise, in accordance with procedures approved by the National Research Council's Report Review Committee. The purpose of this independent review is to provide candid and critical comments that will assist the institution in making its published report as sound as possible and to ensure that the report meets institutional standards for objectivity, evidence, and responsiveness to the study charge. The review comments and draft manuscript remain confidential to protect the integrity of the deliberative process. We wish to thank the following individuals for their review of this report:

Genevieve Atwood, Earth Science Education, Salt Lake City, Utah

Peter T. Bobrowsky, Canada Landslide Program, Geological Survey of Canada, Ottawa, Ontario

Steve Briggs, Community Development and Planning Department, Cincinnati, Ohio

Derek H. Cornforth, Cornforth Consultants, Portland, Oregon

Vicki J. Cowart, Planned Parenthood of the Rocky Mountains, Denver, Colorado

David Noe, Colorado Geological Survey, Denver

Raymond A. Price, Department of Geological Sciences and Geological Engineering (*emeritus*), Queen's University, Kingston, Ontario, Canada

Wilson H. Tang, Department of Civil Engineering, The Hong Kong University of Science and Technology, Kowloon

George A. Thompson, Department of Geophysics (*emeritus*), Stanford University, California

Although the reviewers listed above have provided many constructive comments and suggestions, they were not asked to endorse the conclusions or recommendations nor did they see the final draft of the report before its release. The review of this report was overseen by William L. Fisher, Department of Geological Sciences, University of Texas, Austin. Appointed by the National Research Council, he was responsible for ensuring that an independent examination of this report was carried out in accordance with institutional procedures and that all review comments were carefully considered. Responsibility for the final content of this report rests entirely with the authoring committee and the institution.

Contents

Executive Summary

Landslides occur in all geographic regions of the nation in response to a wide variety of natural conditions and triggering processes that include storms, earthquakes, and human activity. Landslides in the United States constitute a serious hazard that cause substantial human and financial losses, estimated to average 25 to 50 deaths annually and to cost approximately $1 billion to $3 billion per year. In addition to direct and indirect financial losses, landslides cause significant environmental damage and societal disruption. Primarily because individual landslides usually affect limited local areas and individual landowners, damage resulting from landslide hazards has not generally been recognized as a problem of national importance and has not been addressed on a national basis. The absence of a coordinated, national approach to mitigating the detrimental effects of landslides has resulted in a reduced ability of state and local government agencies to apply the important lessons learned, often at considerable expense, in other parts of the country.

As a result of a congressional directive, the U.S. Geological Survey (USGS) addressed the need for a national approach by preparing the National Landslide Hazards Mitigation Strategy (Spiker and Gori, 2000).[1] The proposed strategy describes in broad overview the nine major components, ranging from basic research activities to improved public policy measures and enhanced mitigation, considered as the essential elements

---

[1] A modified version of this report, with the same title, was recently published as USGS Circular 1244 (Spiker and Gori, 2003).

required to address the hazards arising from landslides at a national level. The National Research Council was asked to review this proposal, with the charge given in Box ES.1.

The review committee established to address this charge received input from a wide variety of interested parties during its information-gathering meetings—from federal agencies, state agencies, local jurisdictions, private companies, and the academic community. Based on this input and its own collective experience, the committee was particularly cognizant both of the diversity of issues associated with the national landslide problem that arise from regional considerations and of the considerable variations in institutional capability and responsibility at regional and local levels. It is this range of capabilities, and the widespread demand at the local level for tools and information to address this national problem that present such a clear argument for the coordination and assistance that would be provided by a national program for landslide hazards mitigation. **The committee agrees that a national approach to the mitigation of landslide hazards is needed and considers that the nine**

BOX ES.1
STATEMENT OF TASK

In response to a request from the U.S. Geological Survey, an ad hoc committee established under the auspices of the Board on Earth Sciences and Resources will provide advice regarding the optimum approaches and strategies that could be applied to implement federal-state-local-private partnerships to mitigate the effects of landslides and other ground failures. The study committee will:

• Assess the approach described in USGS Open-file Report 00-450, National Landslide Hazards Mitigation Strategy, comment on the federal-state-local-private partnership concept described in that report, and evaluate whether all the appropriate partners that should be involved in a national landslide hazard mitigation strategy are identified. This assessment should be provided in the form of a brief interim report.
• Consider the potential roles for each of the federal, state, local, and private sectors, and provide advice regarding implementation and funding strategies to stimulate productive, effective, coordinated partnerships.

As part of its analysis, the committee will provide an overview of research priorities required to support the activities of each sector.

components briefly described in the USGS proposal are the essential elements of a national landslide hazard mitigation strategy.

Responsibility for the problems posed by landslides, and for the solutions to those problems, is widely shared among different levels of government and among different stakeholders at each level. This shared responsibility emphasizes the role of the partnerships that will be required to develop and implement a national landslide hazards mitigation strategy. A key starting point for considering landslide partnerships is the recognition that for a national policy to be effective, it must shape not only federal actions but also those of state and local governments, and ultimately those of private landowners. **The committee agrees that a national landslide hazards mitigation strategy should be based on partnerships involving federal, state, local, and nongovernmental entities.**

The description of the components of the national landslide hazards mitigation strategy proposed by the USGS is brief and requires a more complete discussion of the comparative importance of each element. The committee concludes that any analysis and discussion of the proposed national strategy should include a sense of priorities, and accordingly, the commentary and recommendations presented in the following paragraphs are designed to convey the committee's priorities for a national program:

The committee recommends that a national strategy for landslide loss reduction promotes the use of risk analysis techniques to guide loss reduction efforts at the state and local level. Because the state of the art of landslide risk analysis is evolving, further development of risk analysis methods, and documentation and dissemination of their use, are important components of the research and application program for a national landslide strategy. Use of risk analysis for guiding appropriate choice of landslide loss reduction tools should be an important element of the technical assistance and outreach programs provided to state, local, and nongovernmental entities.

The National Landslide Hazards Mitigation Program must play a vital role in evaluating methods, setting standards, and advancing procedures and guidelines for landslide hazard maps and assessments. National landslide hazard information gathering and mapping should be undertaken as a component of the proposed partnerships. The program must establish standards and procedures for the collection, long-term management, and maintenance of this information. Metadata[2] must be associated with all data collected under the auspices of the program, in

---

[2]Metadata refers to information about data (e.g., information describing data source, date collected, method of collection, etc.)

accordance with National Spatial Data Infrastructure protocols. Hazard zonation mapping must be developed for multiple mapping scales by utilizing the best available technologies and accurate, high-resolution terrain information.

In order to provide tools for landslide hazard mitigation, it will be necessary to conduct basic research on monitoring techniques and on aspects of landslide process mechanics. An integrated research program is recommended in which intensive field studies are used to (1) improve site and laboratory characterization techniques; (2) develop new field monitoring methods; (3) obtain greater understanding of failure and movement mechanisms; and (4) develop and test models to predict failure timing, location, and ultimate mass displacement. Studies of debris flows, bedrock slides, and submarine landslides deserve greatest attention. Innovative remote-sensing technologies are now offering researchers the possibility of rapid and detailed detection and monitoring of landslides. Additional support to exploit these new technologies and develop practical tools for a broad user community is needed.

Improved education and awareness of landslide hazards and mitigation options, for decision-makers, professionals, and the general public, must be primary components of a national landslide hazard mitigation program. Collecting and disseminating information on landslide hazards to federal, state, and local governmental agencies and nongovernmental organizations, planners, policy makers, and private citizens in a form useful for planning and decision making is critically important for an effective mitigation program. Education and awareness partnerships will be most effective if implemented at the outset of the program. If the national landslide hazard mitigation program is to materialize, then broad-based acceptance, participation, and support are essential for its success.

The committee agrees that substantially increased funding will be necessary to implement a national landslide hazards mitigation program. The committee considers that the figure of $20 million, presented in the USGS proposal as the amount required to support an enlarged Landslides Hazards Program within the USGS, would provide an adequate basis for the initial stages of a national strategy with a 10-year target for achieving substantial loss reduction goals. However, the committee considers that over the course of the program, the distribution of funds should progress from an initial emphasis on research, development of guidelines, and start-up to the later widespread implementation of landslide risk reduction measures through various partnership programs. The committee considers that additional increases—to annual funding of $35 million for years 4-6 and $50 million for years 7-10 and beyond—will be required to support these later parts of the program. The committee recognizes the reality that national budgetary considerations will determine the total annual funding provided to implement the strategy and emphasizes that it is the distribution of total available funding among the program's different components that is of paramount importance for an appropriately balanced national program.

The committee commends the USGS for undertaking the important initial steps toward a comprehensive national landslide hazards mitigation strategy. The committee recommends that the USGS—in close partnership with other relevant agencies—produce the implementation and management plans that will provide the practical basis for an effective national strategy that can be applied at the local level.

1

Introduction

The surface of the land is made by Nature to decay . . .
Our fertile plains are formed from the ruins of the mountains.
James Hutton, 1785

The surface of the earth, both on land and beneath the oceans, is continually being modified by mass movements operating in response to gravitational forces. One effect of the mass movements termed landslides can be to reduce the gradient of hillslopes to stable angles. In this report, the term "landslide" will include all types of gravity-caused mass movements, ranging from rock falls, through a variety of slumps and slides, to debris flows. Both subaerial and submarine mass movements are included. Although precipitation, earthquakes, and volcanic eruptions are the principal natural drivers of landslides, in many cases landslides result directly from disturbance of hillsides by road construction or other human activity.

Landslides contribute to the erosion, transport, and deposition of earth materials. Over geologic time, they help produce stable land suitable for agriculture and habitation and provide materials that form fertile plains and valleys, beaches, and barrier islands. Unfortunately, landslides are not completely benign to human beings, and because at the scale of the typical human life span the benefits accruing from landslides are overshadowed by their destructive characteristics, they are viewed as hazards that should be understood and, if possible, mitigated. This report is focused on the identification, understanding, and mitigation of landslide hazards—the destructive aspect of landslides.

1.1 PROPOSAL FOR A NATIONAL STRATEGY

The Disaster Relief Act of 1974 (now the Robert T. Stafford Disaster Relief and Emergency Assistance Act—the Stafford Act) assigned responsibility for landslide hazard warning to the Director of the United States Geological Survey (USGS), providing a basis for the USGS to assume a prominent leadership role in national landslide hazard mitigation. The primary objective of the existing USGS Landslide Hazards Program is to reduce long-term losses from landslide hazards by improving scientific understanding of the causes of ground failure and suggesting mitigation strategies. The USGS Landslide Hazards Program has hitherto been funded at a modest level of $2 to $3 million each year. However, impetus for an increased emphasis on this program was provided by the House Report accompanying the Department of the Interior Appropriations Bill for FY 2000, which directed the USGS to develop a comprehensive strategy to address the hazards posed by landslides. During 1999-2000 the USGS convened a series of workshops and meetings to plan and develop a national strategy, resulting in the compilation of USGS Open-File Report 00-450, National Landslide Hazards Mitigation Strategy—A Framework for Loss Reduction (Spiker and Gori, 2000).[1] This report proposed a national strategy based on partnerships between the USGS—as the responsible federal agency—and an array of federal, state, local, community, and industry partners. This partnership strategy envisioned a substantially increased federal investment for the USGS Landslide Hazards Program, requiring almost an order-of-magnitude increase from the present annual funding level of $2.6 million to at least $20 million. Of this total, $10 million would support increased USGS activities and $10 million would be provided to partners.

The USGS strategy proposal (Spiker and Gori, 2000) presents an outline of the elements required for a national approach to the landslide hazard problem, with the 10-year goal of reducing the risk of loss of life, injuries, economic costs, and destruction of natural and cultural resources caused by landslides. The report identifies nine elements of a national landslide hazard mitigation program: (1) research to develop a predictive understanding of landslide processes; (2) hazard mapping to delineate susceptible areas; (3) real-time monitoring of active landslides; (4) loss assessment to determine economic impacts of landslide hazards; (5) information collection, interpretation, and dissemination to provide an effective system for information transfer; (6) guidelines and training for scientists, engineers, and decision makers; (7) public awareness and edu-

---

[1]A modified version of this report, with the same title, was recently published as USGS Circular 1244 (Spiker and Gori, 2003).

cation; (8) implementation of loss reduction measures; and (9) emergency preparedness, response, and recovery to build resilient communities.

The partnerships referred to in the USGS strategy document (Spiker and Gori, 2000) are described in broad outline:

- Partnerships with state and local governments to assess and map landslide hazards, to be funded through competitive grants ($8 million annual allocation, requiring 30% matching funds)
- Partnerships with other federal agencies (e.g., National Park Service, U.S. Forest Service [USFS], Bureau of Land Management) to increase the capabilities of federal agencies to address landslide hazards ($2 million for USGS participation as requested by other agencies)
- Partnerships with universities, local governments, and the private sector to support research and implementation efforts ($2 million annually, distributed through competitive grants)

The committee has reviewed the National Landslide Hazards Mitigation Strategy (Spiker and Gori, 2000) and agrees that the nine major components identified in the proposed national strategy, ranging from basic research activities to improved public policy measures and enhanced mitigation, are the essential elements required to address the hazards arising from landslides at a national level. However, the treatment of these components in the strategic plan is brief and requires a more complete description of the comparative importance of each element. The committee considers that in its analysis and discussion of each element of the proposed national strategy, it is essential that a sense of priorities be presented.

In framing its assessment and review of the National Landslide Hazards Mitigation Strategy, the committee has been particularly aware of the diversity of issues associated with the national landslide problem that arise from regional considerations and of the considerable variations in institutional capability and responsibility at the regional level. It is this range of capabilities, and the widespread demand at the local level for tools and information to address this national problem, that present such a clear argument for the coordination and assistance that would be provided by a national program for landslide hazards mitigation.

1.2 COMMITTEE CHARGE AND SCOPE OF STUDY

To be assured that the strategy advanced by the USGS was the most appropriate approach to this problem, the USGS requested that the National Research Council (NRC) conduct a review, with the charge presented in Box 1.1.

BOX 1.1
STATEMENT OF TASK

In response to a request from the U.S. Geological Survey, an ad hoc committee established under the auspices of the Board on Earth Sciences and Resources will provide advice regarding the optimum approaches and strategies that could be applied to implement federal-state-local-private partnerships to mitigate the effects of landslides and other ground failures. The study committee will:

• Assess the approach described in USGS Open-file Report 00-450, National Landslide Hazards Mitigation Strategy, comment on the federal-state-local-private partnership concept described in that report, and evaluate whether all the appropriate partners that should be involved in a national landslide hazard mitigation strategy are identified. This assessment should be provided in the form of a brief interim report.
• Consider the potential roles for each of the federal, state, local, and private sectors, and provide advice regarding implementation and funding strategies to stimulate productive, effective, coordinated partnerships.

As part of its analysis, the committee will provide an overview of research priorities required to support the activities of each sector.

The review committee established by the NRC to address this charge received input from a variety of interested parties during its information-gathering meetings—from representatives of federal agencies, state agencies, local jurisdictions, private companies, and the academic community. Participants in these meetings uniformly supported a national approach to providing assistance for state and local agency landslide mitigation efforts. However, the wide variation in the nature and extent of existing state and local agency activities means that these are treated in a more generic sense when compared with federal agencies, where the committee was able to evaluate current activities on a nationwide basis and suggest specific roles in a future national partnership strategy.

One issue that the committee grappled with was the extent to which other—non-landslide—ground failure hazard mitigation should or could be addressed in this assessment. The USGS national strategy proposal addresses only ". . . landslides, the most critical ground failure hazard facing most regions of the nation" (Spiker and Gori, 2000, p. 3) but asserts that the strategy "provides a framework that can be applied to other ground failure hazards." In the absence of any detailed consideration of non-landslide ground failure in the national strategy proposal, the committee considered that it was not possible to provide a review of non-landslide hazard mitigation and, accordingly, has restricted its assessment to landslides. However, the commentary on approaches to mitigation presented in this report, particularly the emphasis on risk-based approaches, applies to ground failure hazard mitigation in the broadest sense.

1.3 SOCIOECONOMIC IMPACTS OF LANDSLIDES

Landslides within the United States constitute a major geologic hazard, occurring in all 50 states and causing on average some 25 to 50 fatalities and damage of approximately $1 billion to $3 billion each year (NRC, 1985; Schuster and Highland, 2001). The socioeconomic effects from the thousands of landslides that occur each year impact people, their homes and possessions; industrial establishments; and transportation, energy, and communication lifelines (e.g., highways, railways, communications cables). Socioeconomic losses are increasing as the pressure of expanding populations causes the built environment to expand into more unstable hillside areas.

Landslides are responsible for considerably greater economic losses and human casualties than is generally recognized—although they represent a significant element of many major disasters, the magnitude of their effects is often overlooked by the news media. The losses attributed to most individual landslides are relatively small, although they can be devastating to individual property owners. Because damage costs are borne mostly by individuals, with only some involvement of federal, state, and local government relief and rehabilitation programs, the nation has largely ignored the financial risk posed by landslides.

Landslide costs include both direct and indirect losses affecting private and public properties. Direct costs can be defined as the costs of replacement, rebuilding, repair, or maintenance resulting from direct landslide-caused damage and destruction of property or installations (Schuster and Fleming, 1986; Schuster, 1996; Schuster and Highland, 2001). All other costs of landslides are indirect, for example:

- reduced real estate values in areas threatened by landslides;
- loss of tax revenues on properties devalued as a result of landslides;
- loss of industrial, agricultural, and forest productivity, and of tourist revenues, as a result of damage to land or facilities or interruption of transportation systems;

- loss of human or domestic animal productivity because of death, injury, or psychological trauma; and
- costs of measures to prevent or mitigate potential landslide activity.

Private costs to individuals or corporations are incurred mainly as landslide-caused damage to land and structures, including private property and corporate industrial facilities. A destructive landslide can result in financial ruin for property owners because landslide insurance or other means to offset damage costs usually are not available.

Public costs are those borne by government agencies—national, regional, or local. Probably the largest direct public costs resulting from landslides are for the repair or relocation of transportation facilities (e.g., Box 1.2). Based on a survey of state transportation departments, Walkinshaw (1992) found that the average annual direct costs of maintenance and repairs to U.S. highways as a consequence of landslide damage from 1985 to 1990 were nearly $106 million. Although this survey reflected actions only for state and federal highways, which represent about 20% of the entire U.S. road system, these have the larger cut-and-fill structures and are likely to account for the majority of landslide costs. Nevertheless, the survey may have underestimated these costs because many state transportation departments do not maintain detailed records of their landslide-related highway maintenance costs.

Indirect public costs are diverse and include such disparate elements as loss of tax revenues, reduced capacity or capability of lifelines, reduced productivity of government forests, and impacts on the quality of sport fisheries. An interesting comparison of indirect to direct costs of landslides was provided by the 1983 closure of heavily traveled U.S. Highway 50 in California as a result of landslide activity, which prevented tourist access to popular Lake Tahoe. Although highway repairs totaled $3.6 million, the estimated loss of tourist revenues was a staggering $70 million (San Francisco Chronicle, 1983).

Widespread and numerous landslide occurrences caused by storms—many of which were probably related to El Niño—have plagued California for the past 50 years. Exceptional landslide activity occurred in 1951-1952, 1956, 1957-1958, 1961-1962, 1968-1969, 1977-1978, 1979-1980, 1982, 1995, and 1997-1998. As an example, in the six southern counties of California total losses due to landslides caused by heavy winter rainfall in 1979-1980 were estimated at $500 million (Slosson and Krohn, 1982). In January 1982, an intense storm triggered 18,000 debris flows and landslides in the San Francisco Bay area, damaging or destroying about 6,500 homes and 1,000 businesses. The direct costs of these landslides were in excess of $66 million, but in addition, 930 lawsuits and claims in excess $298 million were filed against city and county agencies (Smith, 1982). In

BOX 1.2
THE THISTLE LANDSLIDE, UTAH

The mountain areas of the western United States experienced abnormally high precipitation in 1982-1984, probably related to a strong El Niño influence. Thousands of landslides occurred across the region, including a very large debris slide at Thistle, Utah, in April 1983. The Thistle debris slide formed a lake by damming the Spanish Fork River and severed three major transportation arteries: U.S. Highways 6/50 and 89, and the main transcontinental line of the Denver and Rio Grande Western Railroad. The railroad and Utah highway engineers undertook emergency work to reestablish routes outside the devastated area.

Anderson et al. (1984) estimated that the total direct costs of all landslides that occurred in Utah during the spring of 1983 exceeded $250 million. A subsequent economic analysis by the University of Utah (1984) evaluated both direct and indirect costs of the Thistle landslide and estimated the total direct costs to be $200 million. In addition, numerous indirect costs were reported; most of these involved temporary or permanent closure of highways and railroads to the detriment of local coal, uranium, and petroleum industries, several types of businesses, and tourism. Perhaps the largest losses due to the Thistle slide were incurred by the Denver and Rio Grande Western Railroad, which spent about $40 million, mostly to construct a twin-bore tunnel about 900-m long that bypassed the landslide and lake, while facing an additional $81 million in lost revenue due to the closure of the line during 1983. The total costs (direct and indirect) of this slide were probably on the order of $400 million. Although there were no casualties as a result of the Thistle slide, it ranks as the most economically costly individual landslide in North America, and probably in the world (Schuster and Highland, 2001).

January and March 1995, above-normal rainfall in southern California triggered damaging debris flows, deep-seated landslides, and flooding in Los Angeles and Ventura Counties (Harp et al., 1999). The most notable landslide that occurred at this time was the deep-seated La Conchita landslide (Figure 1.1), which, in combination with a local debris flow, destroyed or badly damaged 11-12 homes in the small town of La Conchita (O'Tousa, 1995). In the late winter and early spring of 1998, heavy rainfall again caused major landslide activity and damage totaling approximately $156 million in the 10-county San Francisco Bay region (Godt and Savage, 1999).

The 1983 Thistle landslide in Utah dammed the Spanish Fork River and severed major transportation arteries, causing direct and indirect costs estimated at close to $400 million.
SOURCE: Schuster and Highland (2001).

Although landslides are common throughout the Appalachian region and New England, the greatest landslide hazard in the eastern United States, at least in terms of financial losses within a fairly restricted area, is from landslides affecting clay-rich soils in Pittsburgh, Pennsylvania, and Cincinnati, Ohio (see Box 1.3). Landslides also occur across the Great Plains and into the mountain areas of the western United States, where weathered shales and other clay-rich rocks occur near the surface, and they are particularly common where there are steep slopes, periodic heavy rains, and vegetation loss following wildfires. Earthquakes and volcanoes

FIGURE 1.1 The 1995 La Conchita landslide, southern California.
SOURCE: Schuster and Highland (2001).

BOX 1.3
LANDSLIDE HAZARDS IN CINCINNATI, OHIO

The severity of landslide problems in the Cincinnati area became widely known after Fleming and Taylor (1980) compiled and compared estimates for the annual per capita cost of landslide damage in three metropolitan areas. They concluded that these costs were $5.80 for Hamilton County (greater Cincinnati), compared with $2.50 for Allegheny County, Pennsylvania (Pittsburgh area), and $1.80 for the nine-county San Francisco Bay region in California.

About one-fifth of Cincinnati and Hamilton County has steep hillsides forming natural green belts between hilltop communities and neighboring valley communities. Although landslides have occurred in the area before the 1850s, landslide damage has become increasingly expensive as urban development encroached upon these hillsides. Most landslides in the Cincinnati area occur in surficial materials, most frequently in the colluvium-covered valley sides; landslides involving bedrock failures are extremely rare. The colluvium forms wedge-shaped deposits that are thin on the steeper, higher slopes and gradually become thicker (as much as 45 feet thick) near valley bottoms.

continued

The proximity of housing development and transportation corridors (the Columbia Parkway) to landslide-prone hills is strikingly apparent in Cincinnati, Ohio. Each year this hillside causes problems during spring rains.
SOURCE: File photo, Cincinnati Department of Transportation and Engineering.

BOX 1.3 Continued

The City of Cincinnati and Hamilton County, in conjunction with the general public and the geotechnical community, have undertaken and financed a major effort to reduce the damages from landslides, based on collaborative investigations conducted by federal and state agency geologists, aided by faculty and students at the University of Cincinnati and by private consultants. Federal, state, and local governments, as well as private foundations, have financed these studies. Private individuals have donated additional time and services to community planning groups. The most visible successes are in the quality of new construction, where grading is performed under strict regulations, with inspection of plans and construction activities by technical staff from the city or county. The enthusiastic involvement of both the public and the private sectors is achieving major gains in landslide damage reduction.

also cause landslides: the 1994 Northridge Earthquake in the San Fernando Valley triggered thousands of landslides in the Santa Susana Mountains north of the epicenter.

In terms of loss of life, by far the most disastrous landslides to occur within the United States and its territories have been caused by hurricanes or tropical storms making landfall from the western Atlantic Ocean. Hurricane Camille in 1969 caused extensive debris flows in central Virginia—although the exact number cannot be ascertained; most of the 150 who died as a result of Hurricane Camille are thought to have been victims of debris flows triggered by heavy rains associated with the hurricane (Williams and Guy, 1973). In October 1985, heavy rain in Puerto Rico from Tropical Storm Isabel caused a major rock slide that obliterated much of the Mameyes district of the city of Ponce. The slide destroyed about 120 houses and killed at least 129 people—the greatest death toll in North American history from a single landslide (Jibson, 1992).

1.4 ENVIRONMENTAL CONSEQUENCES OF LANDSLIDES

Although landslides routinely cause local changes to the land surface, occasionally much larger changes are produced that affect subaerial and submarine landscapes, natural forests and grasslands, the quality of streams and other bodies of water, and the habitats of native fauna, both on the earth's exposed surface and in its streams and oceans (Schuster, 2001; Schuster and Highland, 2001). Giant prehistoric landslides have been identified on the slope of Sinking Creek Mountain in the Appalachians of Virginia. These are among the largest known landslides in eastern North America and among the largest in the world. They were probably triggered in colder and wetter conditions during the last Ice Age, and these areas appear stable today. However, they have produced many disrupted terrain features, including scarps, ponds, and wetlands, that form important local ecosystems within the National Forest. Although historic landslide events usually are not as large as these examples, many large landslides have had significant impact on topography. One example is the Thistle, Utah, landslide of 1983 that dammed the Spanish Fork River, causing the valley to be flooded and forcing the relocation of major transportation routes (see Box 1.2).

High volumes of landslide-derived sediment can be delivered to stream channels. Debris flows can follow stream channels for great distances, causing substantial channel modification, and they also provide important sediment transport links between hillslopes and alluvial channels. Two examples illustrate these points. Studies in northern Idaho show that rotational landslides produce about 40% of stream sediment, debris avalanches produce an additional 40%, and only 20% is derived from surface-flow erosion (Wilson et al., 1982). In a similar study in Puerto Rico, Larsen and Torres Sanchez (1992) found that 81% of sediment transported within the Mameyes River basin was contributed by mass wasting. An equally important aspect is that sediment levels may remain high for decades following major landslide events, increasing flood risks for downstream communities and threatening efforts to restore fisheries and aquatic ecosystems (e.g., Madej, 1995). Nearly 20 years after the eruption of Mount St. Helens, sediment eroded from landslide and debris flow deposits continues to elevate sediment levels in the Toutle River by 10 to 100 times (Bernton, 2000).

Although most kinds of wildlife are able to retreat fast enough to prevent injury from all but the fastest-moving landslides, all wild creatures are subject to landslide-caused habitat damage and destruction. Birds and other animals that nest or live in underground burrows are at high risk. Schuster and Highland (2001) reported that springtime landslides in fluvial-lacustrine sediments along the Columbia River in south-central Washington probably kill very large numbers of nesting cliff swallows.

Landslides can adversely impact fish habitats, especially those of anadromous fish (e.g., salmon) which live in the oceans but return to freshwater streams to spawn. Landslides cause many changes in aquatic habitat. Elevated sediment delivery due to landslides can lead to increased mobility and scour of spawning gravels, increased fine sediment in spawning gravels, increased fine sediment in pools, bank destabilization, and diminished availability of food organisms (e.g. Swanston, 1991). In the Pacific Northwest, the increased occurrence of debris flows due to timber harvest activities (Sidle et al., 1984; Schmidt et al., 2001) and the coincident rapid decline in aquatic habitat and salmon stocks have led to the enactment of numerous state and federal land-use regulations and to frequent lawsuits over their application and validity (e.g., see University of California Committee on Cumulative Watershed Effects, 2001). The USFS has conducted numerous studies of the relationship between destructive landslides, forest cover, and logging operations in this area (Swanston and Swanson, 1976; Megahan et al., 1978; Swanston, 1991; McClelland et al., 1999). It has become standard practice in assessing timber management plans to conduct watershed-scale surveys to map landslides (and landslide potential) and their possible association with and sensitivity to land-use practices. There is debate as to the most efficient and accurate way to do this, and how to interpret such maps in terms of land-use decisions. No standards have been established in the United States, and maps vary widely in their detail and accuracy and, consequently, their usefulness. It is important that not only should landslide occurrence or relative slope stability be established for an area, but also the sediment and wood production associated with landslides should be estimated, because sediment and woody debris strongly influence aquatic habitats. Federal leadership is needed to establish methods and standards of reporting and interpretation.

Landslides have occasionally directly caused human health problems. Following the 1994 Northridge, California, earthquake, Ventura County experienced a major outbreak of coccidioidomycosis (valley fever)—a respiratory disease contracted by inhaling airborne fungal spores (Jibson et al., 1998). The earthquake and its aftershocks produced many highly disrupted, dust-generating landslides in canyons northeast of Simi Valley, and prevailing winds transported dust into the Simi Valley and to communities farther west (Figure 1.2). In the following eight weeks, 203 coccidioidomycosis cases were reported, about an order of magnitude more than would otherwise be expected. The temporal and spatial distribution of coccidioidomycosis cases indicated that the outbreak resulted from inhalation of spore-contaminated dust generated by the earthquake-triggered landslides.

1.5 THE CONCEPT OF LANDSLIDE MITIGATION

The preceding sections, and the suite of case studies presented in Appendix A, highlight the immense socioeconomic and environmental damage and losses that result from landslides. A considerable variety of

FIGURE 1.2 Dust clouds created by earthquake-triggered landslides in Santa Susana Mountains. Dust was blown southwestward over Simi Valley where an epidemic of valley fever (coccidioidomycosis) occurred.
SOURCE: Harp and Jibson (1995).

techniques and practices have been employed to mitigate the potential losses arising from landslide hazards. Although landslide losses will be negligible if all landslide hazards are correctly identified and landslide-prone areas are avoided, such an approach is rarely feasible, and it is neither possible nor desirable to proscribe development on all landslide-prone areas. The question then becomes one of identifying the most effective of the various mitigation approaches and obtaining funding to apply the optimum mitigation.

The USGS proposal for a national landslide hazards mitigation strategy (Spiker and Gori, 2000) clearly summarizes the major mitigation approaches, including

- restricting development in landslide-prone areas;
- enforcing codes for excavation, construction, and grading;
- engineering for slope stability;
- deploying monitoring and warning systems; and
- providing landslide insurance.

This emphasizes that mitigation strategies extend far beyond engineering or technical measures designed to stabilize a landslide. Achieving mitigation by enforcing strict building code, excavation, and fill ordinances, as applied in the Cincinnati area (see Box 1.3), depends upon a sound scientific knowledge of the distribution and properties of the earth materials and potential landslide processes. Local zoning and subdivision regulations in California have been used creatively by some communities to concentrate building on stable land while leaving unstable land as open space to serve the development. This model could be adapted in other states. Also in California, public districts have been formed to provide funds for potential landslide repairs. Although mitigation options could include this type of financial arrangement to pool the exposure of individuals to landslide hazards and to compensate for losses, such mechanisms have not been widely applied in the United States and there seem to be considerable obstacles to providing a broadly based national landslide insurance program (see section 5.2). As a consequence, the mitigation of landslide hazards for many existing developments and transportation networks is accomplished by using a variety of monitoring and warning systems—which protect lives and property but do not prevent landslides—or by resorting to expensive engineering stabilization solutions.

Engineering solutions are often expensive and may not provide a permanent solution. In some cases a "permanent" solution is unnecessary—the landslide hazard is high for only a relatively short time, such as during excavations for a new structure. One example of early and innovative temporary landslide hazard mitigation is provided by the construction of Grand Coulee Dam on the Columbia River in Washington State (U.S. Bureau of Reclamation, 1936; Hansen, 1989). The north abutment of the dam was threatened when a huge mass of water-soaked silt began to creep into the excavation. Engineers placed numerous pipes into the silt mass and a refrigerant was pumped through these pipes, thereby freezing the water in the silt and temporarily stabilizing the landslide until the concrete for the abutment had been poured and cured. This temporary mitigation activity was accomplished in about six weeks, saving considerable time and expense.

Occasionally, opportunities arise that allow for ingenious engineering solutions of a more permanent nature. The construction of Interstate 70 over Vail Pass in the 1970s encountered highly unstable conditions and numerous landslides along the route, especially on the western side of Vail Pass where several sections of almost continuous landslides are found along Black Gore Creek. Many innovative concepts and designs were utilized or developed on this project because slope stability and erosion control were paramount concerns. At one location on the western side of Vail Pass, two landslides on opposite sides of the valley were stabilized by allowing them to buttress each other. Fill was added to the valley, the stream profile was raised and controlled to prevent further erosion of their toes, and the highway was constructed on the fill (Robinson and Cochran, 1983). Where land values are high or the need for a lifeline is high, the cost of expensive engineering solutions can be justified. In other instances, especially in larger developments, it may be preferable to build on stable terrain and use the potentially unstable terrain as valuable open space.

One can argue that the most beautiful parts of the world are where the hazard of landslides is greatest and, because such areas are increasingly a focus for development, where the risks are most extreme. For this reason, the present report emphasizes the identification and mitigation of landslide hazards rather than either completely disallowing development or proposing to prevent landslides. By knowing the hazard and the risk, the informed citizen, developer, or public official can calculate the trade-offs among beauty, health, and wealth, and between fortune and fate. We must learn to live intelligently among the mountains and their ruins.

1.6 OVERVIEW OF NATIONAL STRATEGY PRIORITIES

Landslides are widely distributed geographically and pose differing types of hazards depending on geologic setting and terrain type. The diversity of landslide problems, and the breadth of the needed elements of a national landslide hazard reduction program, can be illustrated by examples:

- Debris flows triggered by extreme rainfall events have had devastating effects in mountainous regions of the United States, and there are indications that differences in climate, materials, vegetation, and topography may cause a variety of debris flow phenomena in different regions. Accordingly, debris flow hazard reduction will require improved understanding of the initiation and propagation of these flow events and their region-specific characteristics. Once such basic science questions have been answered, the hazards posed by debris flows can be reduced by application of appropriate risk assessment and mitigation techniques.
- Rock falls pose severe hazards, particularly along transportation corridors in many mountainous states. Although often involving only a small volume of material, the speed of rock falls and the hazard they pose to motorists have prompted several state departments of transportation to support rock fall research. The science of rock fall mechanisms is relatively well understood, and several computer simulation programs have been developed to aid in evaluating the hazard. However, improvements are needed in establishing standards for risk management and for certify-

ing the effectiveness of rock fall barrier systems. This can be achieved by encouraging more widespread adoption of established techniques through technology transfer.

• Bedrock slides occur in many locations throughout the United States, displaying a range of movement types and resulting from a wide variety of triggering events. They can be local events involving small volumes or multiple events and large masses that involve a considerable area. Slides can move quite slowly or very rapidly, and the hazards they pose range from relatively minor to catastrophic. Once initial movement has occurred, bedrock slides can be mapped readily if the needed resources are available. Nevertheless, they continue to cause extensive economic losses due to ineffective regulatory controls on development in slide-prone areas.

In general, improved risk assessment is needed for all types of landslide hazards (see Chapter 4 below), as are advances in methods of cost-effective mitigation that might include hazard insurance and other financial instruments. Specifically, the establishment of landslide hazard mitigation priorities should incorporate existing knowledge and the potential for cost-effective results.

The matrix presented in Figure 1.3 evaluates the six broad landslide types[2] against five activities that should be included in an effective national strategy to address the diversity of landslide hazard problems:

1. *Improvement of the science base* to provide an adequate understanding of landslide triggering and landslide movement mechanisms is an essential first step to fill gaps in current understanding and is a fundamental requirement for other activities.

2. *Technology integration and transfer* is important for both the dissemination of scientific understanding of the hazard and the identification of appropriate mitigation methods.

3. *Mapping and monitoring* provides the fundamental database for identification and delineation of landslide hazards.

4. *Risk assessment* integrates the many factors relating to slide occurrence and consequence; it can be applied at various levels, ranging from qualitative to quantitative.

5. *Mitigation* takes many forms, with land-use regulation being the most important. Other mitigation activities include stabilization through engineering activities and construction of diversion works.

---

[2]A detailed classification and description of the numerous landslide types is presented in Cruden and Varnes (1996); these have been simplified into six broad landslide categories for this report.

Activity / Type	Improve Science Base	Technology Transfer	Mapping & Monitoring	Risk Assessment	Mitigation
Debris Flow					
Rock Fall					
Bedrock Slide					
Liquefaction Flow					
Soft Clay Slides					
Submarine Landslides					

FIGURE 1.3 Matrix evaluating the six broad landslides types against the activities required for an effective national strategy. Shaded boxes indicate activities with the highest payoff potential.

This matrix emphasizes that for a range of reasons described throughout this report, some of the component activities have greater "payoff" potential than others and should be accorded greater priority in a national strategy. This variability of potential depends on the particular type of landslide:

• *Debris Flow*: Investment in basic research to improve understanding of debris flow initiation and movement has a high payoff potential. Although scientific understanding of the phenomenon is still inadequate to identify technology transfer as having the highest payoff potential, much could be achieved with better risk-based zonation. The basic scientific advances will also contribute to improved mapping, which is a priority requirement. In addition, clarification of magnitude-frequency-runout characteristics can be anticipated, and these are important for risk assessment and mitigation (including regulation). Improved mitigation methods and the establishment of appropriate risk assessment techniques are needed.

- *Rock Fall*: Rock fall processes are relatively simple and reasonably well understood. The Federal Highway Administration and some state highway departments have made substantial progress in developing rock fall hazard rating systems and in technology integration and transfer. It appears that widespread dissemination of this information would encourage implementation and have a high payoff potential. At the same time, improved mitigation methods, such as improved criteria for testing, certifying, selecting, designing, and installing rock fall barriers, are needed. Establishment and broad acceptance of appropriate risk assessment techniques are also required.

- *Bedrock Slides*: There is reasonable understanding of the mechanics of bedrock slide initiation, although additional case histories would add significantly to the body of knowledge. Although rare, very large, fast-moving bedrock slides are potentially life-threatening, and their movement dynamics are poorly understood and require further study. Post-failure movement dynamics and deformations of many bedrock slide types are poorly understood, and further expansion of the science base in this area is desirable. Once initial movement has occurred, bedrock slides can be identified with current technology and there is high payoff potential associated with mapping them in areas of high risk in order to assist regulation. Improved mitigation methods and the establishment of appropriate risk assessment techniques are needed.

- *Liquefaction Flow*: Liquefaction flows are often caused by earthquakes, but they can also occur in some types of glaciomarine clays. The basic science of liquefaction flow has received considerable attention in recent years, and liquefaction susceptibility criteria have been established and tested in the field. A high payoff potential can be expected from mapping this hazard. As with other landslide types, improved mitigation methods and the establishment of appropriate risk assessment techniques are needed.

- *Soft Clay Slides*: Geotechnical engineers have devoted substantial effort to understanding the mechanics of soft clays; as a result, the initiation and movement of landslides in these deposits are the best understood of all landslide types. Mapping is straightforward. Improved mitigation methods and the establishment of appropriate risk assessment techniques are needed.

- *Submarine Landslides*: Because of the likelihood that submarine landslides will cause highly destructive tsunamis (e.g., the 1998 Papua New Guinea tsunami; Bardet et al., 2003; Liam Finn, 2003; Wright and Rathje, 2003), there is an urgent need to better understand the mechanics of submarine slide movement, particularly the role of gas hydrates in causing shelf edge instability. At present there is a vast amount of geotechnical data scattered among hydrocarbon exploration and development companies, offshore geotechnical companies, and academic institutes, and these data have to be collated and the gaps identified. Areas of potential submarine failure will have to be mapped and procedures determined for submarine landslide risk assessment.

The elements of a national landslides hazards mitigation strategy are dealt with in more detail in the following chapters; Chapter 2 describes requirements and priorities for research into landslide processes; Chapter 3 describes the status of mapping and monitoring techniques and their application; Chapter 4 describes the importance of loss and risk assessment; Chapters 5 and 6 describe the technology transfer and integration components of a national mitigation strategy; and Chapters 7 and 8 describe the partnerships and funding that will be required for implementation of an effective national strategy. Chapter 9 contains the committee's conclusions and recommendations.

2

Research Priorities in Landslide Science

Although the Disaster Relief Act of 1974 (now the Robert T. Stafford Disaster Relief and Emergency Assistance Act—the Stafford Act) delegates the responsibility to issue disaster warnings for landslides to the Director of the U.S. Geological Survey (USGS), it is important to appreciate that this is neither an easy nor a routine task. Some landslide events are widespread, whereas others are local. Some occur suddenly, while others develop slowly over time. Many landslides are triggered by ground saturation caused by intense storms, spring snowmelt, or irrigation and other human disturbance of surface or subsurface drainage systems. Others are triggered by earthquakes and volcanoes, and still others appear to occur for no obvious reasons. Therefore, an understanding of landslide processes—the science of landslides—is an essential requirement both for issuing warnings and for undertaking the host of other mitigation activities ranging from land-use planning to the construction of engineered solutions.

The National Landslide Hazards Mitigation Strategy (Spiker and Gori, 2000) proposes that the USGS should lead a research program directed at developing a predictive understanding of landslide processes and triggering mechanisms. The strategic objective of such a research program would be the following:

- Develop a research agenda and an implementation plan to improve understanding of landslide processes, thresholds, and triggers and to improve the ability to predict landslide hazard behavior.

26

- Develop improved scientific models of ground deformation and slope failure processes that could be implemented in predicting landslide hazards.
- Develop predictive systems capable of interactively displaying changing landslide hazards in both space and time in areas prone to different types of hazard-triggering mechanisms, such as severe storms and earthquakes.

The committee concurs that an expanded research effort that would contribute to an improved understanding of landslide processes and their triggering is an essential component of a national landslide hazards mitigation program. However, such research activities should be prioritized to address those areas of landslide science with the highest payoff potential—namely, debris flow, bedrock slide, and submarine landslide mechanisms (as outlined in section 1.6). This chapter focuses on the strategic research objectives presented in the proposal and an assessment of the role and efficacy of such research in landslide hazard mitigation.

Reliable landslide warnings and effective mitigation must be underpinned by an understanding of the mechanics of landslide processes. For any potential landslide situation, this implies finding answers to the following questions:

- How would the landslide be initiated?
- What are the warning signs?
- How large will it be?
- How far will it move?
- How fast will it move?

Answers to the preceding questions will vary with landslide type and with the nature of the material composing the slide mass. A single rock landing on a highway may cause dire results, and a small, fast-moving landslide in a high-population-density area may pose a greater threat to public safety than a large, slow-moving slide. In Hong Kong, slides with volumes as low as 200m^3 have caused fatalities (Works Bureau, 1998).

The diversity of landslide problems was emphasized in the matrix presented in Figure 1.3, showing those activities that have a high payoff potential within the next five years. In addition, although many aspects of landslide process mechanics are well understood for many landslide types, the understanding of slide mobility—the mass and speed of earth movements—is inadequate to support hazard warnings and other means of mitigation for *all* landslide types. The scientific research program for a national landslide hazards mitigation strategy should include investigating models of ground deformation, slope failure processes, landslide triggering, and prediction. Basic science activities should be directed toward answering a series of questions:

1. *Debris Flows*: Often triggered by extreme rainfall events, debris flows have had devastating effects in mountainous regions.

- How are they initiated?
- How can runout characteristics be established?
- What controls their magnitude-frequency relationships and return periods?

2. *Bedrock Slides*: A large part of the United States is underlain by weak bedrock in which ancient and current landslides are found.

- What factors control the distribution of bedrock slides?
- How do they respond to climatic events?
- What controls their velocity?
- How safe should a stable slide be to support development?

3. *Submarine Slides*: The national strategy for landslide hazard mitigation should extend to the offshore, recognizing the special problems—particularly the difficulties involved with surface and subsurface sampling and in situ geotechnical measurement—associated with submarine landslides.

- How can they be effectively mapped?
- What is the role of gas hydrates in slope instability?
- How can their geotechnical characteristics be assessed?
- How can geotechnical characteristics be translated into risk assessments?

To answer these questions, a comprehensive research program should be designed to produce improvements in the following:

- in situ characterization,
- laboratory characterization,
- advances in formulating geomechanical and geohydrological models,
- advances in kinematic modeling, and
- field studies at sites to facilitate in situ characterization and model validation.

Many important questions related to landslide processes can be addressed only by a scientific research program based at a number of long-term field sites around the nation, selected for their generic interest and their capacity to yield important results. These problems would range from material characterization, to pore pressure response studies, to geomechanical analyses, to large deformation evaluation, according to the specific priorities for landslide type and geological environment. The selection of field sites and the development of site-specific programs should be based on partnerships between the USGS and other federal or state agencies. Specific activities at a given site could be undertaken by the private sector as well as by public agencies, which will facilitate technology transfer. Although priority should be given to debris flow and weak bedrock slide field sites, and to submarine slide sites where possible, other types of slides should not be excluded if a suitable opportunity is available. In the past, landslide activities within the USGS have focused primarily on field-based hazard mapping and assessment. Although mechanistic studies have not been a dominant part of the USGS program, the development of a debris flow flume and related studies into the fundamentals of debris flow mechanics (Iverson, 1997) is an important exception.

In developing a research program, it is important to realize that research into landslide mechanisms by or on behalf of federal agencies has not been the prerogative of the USGS alone. Different agencies bring different skills and experience to address the research agenda:

• In the past, the U.S. Army Corps of Engineers (USACE) conducted extensive research into landslides in clay-shale slopes, particularly when the USACE was closely involved with operation of the Panama Canal (Lutton et al., 1979) and during the construction, operation, and maintenance of major dams on the Missouri River (e.g., USACE, 1983, 1998). In addition, the USACE has both conducted and supported research into seismically induced liquefaction. Although landslide-related research within the USACE is currently at a low level, experimental facilities and experienced personnel exist within laboratories and district offices of that organization.

• Although current expenditures directed toward landslide issues are modest, research into landslide mechanisms has been conducted at universities for many decades, mainly in departments of civil engineering but also within geological engineering and engineering geology programs. Almost all theoretical methods of slope stability analysis have emerged from university-based research.

• Demonstration projects and technology synthesis related to landslide hazard assessment and mitigation have also been supported by the Federal Highway Administration, often in partnership with state departments of transportation. The focus on user needs has been both appropriate and effective. This effort has been directed to a large degree at problems of rock fall and small slides, where the basic scientific understanding is generally adequate. However, there are also situations in which large slides threaten or impact transportation corridors, but because of restricted legislative mandates, the authorities are unable to assess the full range of regional considerations.

• The U.S. Forest Service supports research stations in the Pacific Northwest that have been important in documenting the relationships between timber management practices and slope instability and the influences of landslides on rivers and river ecosystems. It has been the lead agency in establishing the role of root strength in controlling shallow soil stability.

Although the USGS undoubtedly is a major stakeholder in influencing this agenda and in conducting some of the research, the diversity of skills and perspectives of other stakeholders (federal and state agencies, universities, private sector) should be recognized as an asset and incorporated into the national research program at the outset. In particular, the intrinsic value of merit-based competitive selection of research projects, as implemented by the National Science Foundation, should be emphasized as an effective means of conducting such research. Although research into the science of landslide processes, in accordance with the priorities based on payoff potential outlined here, should be undertaken as an important component of a comprehensive national landslide hazards mitigation strategy, the committee emphasizes that such research should be carried out in concert with other critically important research activities—into new technologies for mapping and monitoring; new mitigation approaches; the intermixed physical science and social science issues related to public awareness, understanding, and professional education and capacity building; and particularly, the application of risk analysis techniques to guide mitigation decisions—described in the following chapters.

3

Landslide Mapping and Monitoring

"The identification and map portrayal of areas highly susceptible to damaging landslides are first and necessary steps toward loss-reduction" (Zeizel, 1988). Because individuals or groups do not undertake mitigation actions when they do not understand what to do, lack training, or do not have access to appropriate and understandable technical information, the communication and use of technical information is crucial for effective landslide hazard mitigation programs. There are four general categories of potential users of landslide hazard information (Wold and Jochim, 1989):

1. scientists and engineers who use the information directly;
2. planners and decision makers who consider landslide hazards among other land-use and development criteria;
3. developers, builders, and financial and insuring organizations; and
4. interested citizens, educators, and others with little or no technical experience.

Members of these groups differ widely in the kinds of information they need and in their ability to use that information (Wold and Jochim, 1989).

Most local governments do not have landslide hazard maps and do not have funding available for mapping activities, and such communities usually look to a higher level of government for mapping. The U.S. Geological Survey (USGS) has provided maps in some areas (e.g., demonstration mapping of San Mateo County, California; Brabb et al., 1972), but in general, landslide hazard mapping by the USGS has had limited geographic coverage. Although most local communities look to their state as the primary source of maps, few states have undertaken significant landslide hazard mapping programs. However, there are important exceptions. California and Oregon, for example, have undertaken landslide hazard mapping at standard USGS mapping scales. These maps provide an excellent starting point for local communities and, importantly, form the basis for state laws that require a certain level of compliance with the information they provide.

The considerable variability among state geological agencies, particularly in terms of their existing mapping capabilities and projected funding environments, makes it difficult to provide detailed commentary and suggestions regarding the partnerships between the USGS and states for landslide hazard mapping and assessment. Historically, there have been strong ties between the USGS and state geological surveys in the realm of mapping (e.g., Ellen et al., 1993; Coe et al., 2000b) and, to a lesser extent, for the identification and mitigation of natural hazards. The suggestion in the national strategy proposal (Spiker and Gori, 2000) of mapping partnerships, using a model based on competitive grants and matching funds (as with the existing National Geologic Cooperative Mapping Program), would undoubtedly provide resources for a considerable amount of much-needed mapping. However, such a model raises the possibility that hazard mapping assessed as having a high priority might not be possible if state matching funds are not available. It is important that the details of the cooperative mapping partnership be worked out carefully, in close consultation with state geologists, as the national strategy implementation plan is being developed.

The principles and scope of the landslide hazard mapping, assessment, and delineation task contained within the USGS National Landslide Hazards Mitigation Strategy (Spiker and Gori, 2000) are defined in section 3.1. This section uses several important terms, such as "hazard," "susceptibility," "zonation," and "vulnerability," that are defined in Box 3.1. Landslide hazard zonation is commonly portrayed on maps. Preparation of these maps requires a detailed knowledge of the landslide processes that are or have been active in an area and an understanding of the factors that may lead to an occurrence of potentially damaging landslides. Accordingly, this is a task that should be undertaken by geoscientists. In contrast, vulnerability analysis, which assesses the degree of loss (see Box 3.1), requires detailed knowledge of population density, infrastructure, economic activities, and ecological and water quality values and the effects that a specific landslide would have on these elements. Specialists in urban planning and social geography, economists, and engineers should perform these analyses.

BOX 3.1
DEFINITIONS

The terminology used in this report concerning landslide hazards and associated concepts reflects the following definitions, based on Varnes (1984), the Australian Geomechanics Society (AGS, 2000), and the more general terminology presented in the International Strategy for Disaster Reduction (ISDR) draft report (UN, 2002):

- *Landslide hazard* refers to the potential for occurrence of a damaging landslide within a given area; such damage could include loss of life or injury, property damage, social and economic disruption, or environmental degradation.
- *Landslide susceptibility* refers to the likelihood of a landslide occurring in an area on the basis of local terrain conditions. Susceptibility does not consider the probability of occurrence, which depends also on the recurrence of triggering factors such as rainfall or seismicity. The terms hazard and susceptibility are frequently used incorrectly as synonymous terms.
- *Landslide vulnerability* reflects the extent of potential loss to a given element (or set of elements) within the area affected by the hazard, expressed on a scale of 0 (no loss) to 1 (total loss); vulnerability is shaped by physical, social, economic, and environmental conditions.
- *Landslide risk* refers to the probability of harmful consequences— the expected number of lives lost, persons injured, extent of damage to property or ecologic systems, or disruption of economic activity—within a landslide-prone area. The risk may be individual or societal in scope, resulting from the interaction between the hazard and individual or societal vulnerability.
- *Element at risk* refers to the population, public and private infrastructure, economic activities, ecologic values, et cetera, at risk in a given area.
- *Specific landslide risk* means the expected degree of loss due to a particular landslide, based on risk estimation—the integration of frequency analysis and consequence analysis.
- *Landslide risk evaluation* is the application of analyses and judgments (encompassing physical, social, and economic dimensions of landslide vulnerability) to determine risk management alternatives, which may include determination that the landslide risk is acceptable or tolerable.
- *Landslide hazard zonation* refers to division of the land into homogeneous areas or domains and the ranking of these areas according to their degrees of actual or potential hazard or susceptibility to landslides.

Because landslides both leave a topographic signature when they occur and are driven largely by topographic effects, improved sources of high-resolution topographic information have the potential to greatly increase the accuracy of landslide hazard maps. The probable impacts of new remote-sensing tools on the creation of such maps are reviewed briefly in section 3.2.

The fundamental importance of landslide hazard mapping, assessment, and delineation to the development of effective loss reduction strategies is discussed in section 3.3. It defines the role of landslide zonation mapping in defining priorities for landslide investigations, monitoring activities, or mitigation programs within national, regional, or local landslide hazard mitigation plans.

In some cases, engineered mitigation may be undertaken immediately after the susceptibility of an area is identified, but frequently an engineered approach to mitigation is not cost-effective. In such cases, monitoring systems, discussed in section 3.4, may provide the most appropriate mitigation option. Even in cases where engineered mitigation is eventually planned, interim monitoring capabilities are often installed to ensure that any landslide movement is identified as early as possible so that injury or economic costs can be avoided. Thus, monitoring of landslide-prone regions is an important adjunct to susceptibility and hazard mapping.

3.1 SUSCEPTIBILITY AND HAZARD MAPPING

The USGS proposal for a national strategy (Spiker and Gori, 2000) identified three activities that will be required to provide the maps, assessments, and other information needed by officials and planners to reduce landslide risk and losses:

1. Develop and implement a plan for mapping and assessing landslide and other ground failure hazards nationwide.
2. Develop an inventory of known landslide and other ground failure hazards nationwide.
3. Develop and encourage the use of standards and guidelines for landslide hazard maps and assessments.

The USGS proposal states that "landslide inventory and landslide susceptibility maps are critically needed in landslide prone regions of the nation. These maps must be sufficiently detailed to support mitigation action at the local level. To cope with the many uncertainties involved in landslide hazards, probabilistic methods are being developed to map and assess landslide hazards" (Spiker and Gori, 2000, p.13). The proposal also notes that "these maps and data are not yet available in most areas of the

United States." The committee concurs that the national strategy proposal appropriately identifies the landslide hazard zonation task as the primary responsibility of the USGS, to be undertaken in partnership with states; the following analysis and recommendations concern implementation of this task.

Hazard zonations may be mapped at various scales; user requirements and the intended applications determine the appropriate scale (Box 3.2). Because a clear understanding of the different types of landslide hazard maps is critical for successful implementation of a national strategy, the definitions from Spiker and Gori (2000) are reproduced in Box 3.3. In the absence of accepted national standards for landslide hazard maps, a variety of mapping styles have been employed for each type of map. This even applies to landslide inventory maps—the most basic type of landslide map. These document the locations and outlines of landslides that have occurred in an area during a single event or multiple events. Small-scale landslide inventory maps may show only landslide locations and general outlines of larger landslides, whereas large-scale maps may distinguish landslide sources from landslide deposits, classify different kinds of landslides, and show other pertinent data (Figure 3.1).

The quantitative definition of hazard or vulnerability requires analysis of landslide-triggering factors, such as earthquakes or rainfall, or the application of complex models. Both tasks are extremely difficult when dealing with large areas. Consequently, the legends for most landslide hazard maps usually describe only the *susceptibility* of certain areas to landslides (Figure 3.2), or provide only relative indications of the degree of hazard, such as high, medium, and low.

Over the past decades, geoscientists have developed several approaches to landslide hazard analysis, which can be broadly classified as *inferential, statistical* and *process-based* (Hansen, 1984; Varnes, 1984). All three approaches (Box 3.4) are currently applied to produce the different map types defined in Box 3.3, and there is no standard approach used in the United States. Not all methods of landslide zonation are equally applicable at each scale or for each type of analysis. Some require very detailed input data that can be collected only for small areas because of the required levels of effort and the high cost. Consequently, selection of an appropriate mapping technique depends on the type of landslide problems occurring within an area of interest and the availability of data and financial resources, as well as the duration of the investigation and the professional experience of the experts involved.

When carefully applied by well-qualified experts, the inferential approach may describe the real causes of slope instability, based on scientific and professional criteria. However, due to the scale and complexity of slope instability factors, the basic inferential approach is unlikely to be

BOX 3.2
HAZARD ZONATION SCALES

Characteristics and Use

*Data Acquisition and
Mapping Procedures*

• *National zonation maps:*
Provide a general inventory of
landslide problem areas for the
nation with a low level of detail.
These maps are useful to national
policy makers and the general
public.

National summary of regional
landslide inventories and map
products.

• *Regional zonation maps:*
Provide engineers and planners
an overview of potential landslide
impacts on large projects or
regional developments during
initial planning phases. The areas
investigated are quite large and
the required map detail is low.

Detailed data collection for
individual factors (geomorphology,
lithology, soils, etc.) is not a cost-
effective approach. Data gathered
from stereoscopic satellite imagery
combined with regional geologic,
tectonic, or seismic data should
delineate homogeneous terrain
units.

• *Neighborhood zonation maps:*
Identify landsliding zones for large
engineering structures, roads, and
urban areas. The investigations
may cover quite large areas; yet a
considerably higher level of detail
is required. Slopes adjacent to
landslides should be evaluated
separately and may be assigned
different hazard scores depending
on their characteristics.

Data collection should support
the production of detailed
multitemporal landslide
distribution maps and provide
information about the various
parameters required in statistical
analysis.

• *Site-specific zonation maps:*
Used during site investigations to
provide absolute hazard classes
and variable safety factors as a
function of slope conditions and
the influence of specific triggering
factors.

Data collection should relate to
the parameters needed for slope
stability modeling (e.g., material
sequences and geotechnical
properties, seismic accelerations,
hydrologic data).

BOX 3.3
LANDSLIDE HAZARD MAP TYPES

A **landslide inventory map** shows the locations and outlines of landslides. A landslide inventory is a data set that may represent a single event or multiple events. Small-scale maps may show only landslide locations, whereas large-scale maps may distinguish landslide sources from deposits, classify different kinds of landslides, and show other pertinent data.

A **landslide susceptibility map** ranks slope stability of an area into categories that range from stable to unstable. Susceptibility maps show where landslides may form. Many susceptibility maps use a color scheme that relates warm colors (red, orange, and yellow) to unstable and marginally unstable areas and cool colors (blue and green) to more stable areas.

A **landslide hazard map** indicates the annual probability (likelihood) of landslides occurring throughout an area. An ideal landslide hazard map shows not only the chances that a landslide may form at a particular place, but also the chances that a landslide from farther upslope may strike that place.

A **landslide risk map** shows the expected annual cost of landslide damage throughout an area. Risk maps combine the probability information from a landslide hazard map with an analysis of all possible consequences (property damage, casualties, and loss of service).

---

definitive over large areas when mapping is conducted at small scales. For such applications, the combination of expert inference and qualitatively weighted contributing parameters greatly improves the objectivity and reproducibility of the zonations. Combined statistical and process-based approaches may efficiently provide reliable regional landslide zonations over large areas, by classifying the terrain into susceptibility classes that reflect the presence and intensity of causative factors of slope-instability. For detailed studies of small areas, large amounts of data may become available, and in such cases, simple process-based models become increasingly practical for establishing landslide hazard zonations. They allow variations in the safety factor to be approximated and, thus, yield information useful to design engineers. In an environment where choices must be made from a number of possible mapping approaches, the proposed National Landslide Hazards Mitigation Program can play a vital role in evaluating methods, setting standards, and ratifying procedures.

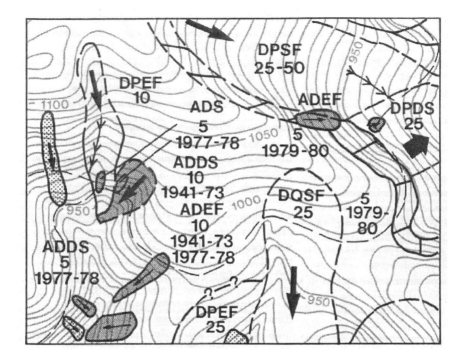

FIGURE 3.1 Detail of inventory map showing recently active and dormant land-slides near La Honda, Central Santa Cruz Mountains, California. Information shown on this map includes state of activity, dominant type of movement, scarp location, and depth and date of movement.
SOURCE: Wieczorek (1982).

3.2 NEW REMOTE-SENSING TECHNOLOGIES

Remote-sensing is used here in its broadest sense to include aerial photography as well as imagery obtained from a variety of platforms, ranging from ground-based mobile units to airborne or satellite platforms. Because landslides directly affect the ground surface, remote-sensing techniques are well suited to slope instability studies. Remote-sensing images can provide diagnostic information concerning the overall terrain conditions that often are critical for determining susceptibility to slope instability. Landslide information extracted from remote-sensing images is related mainly to the morphology, vegetation, and drainage conditions of the slope. The interpretability of slope instability or movements on remote-sensing images is related to both the size of the landslide features and

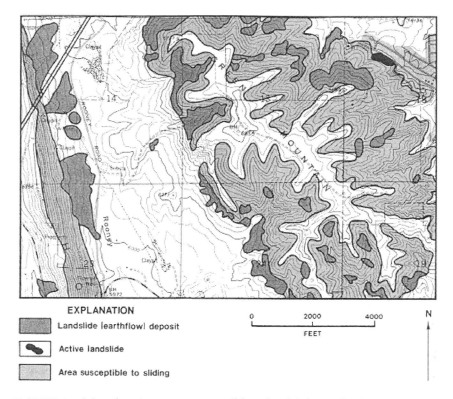

EXPLANATION

Landslide [earthflow] deposit

Active landslide

Area susceptible to sliding

0   2000   4000   N

FEET

FIGURE 3.2 Map showing areas susceptible to landslides in the Green Mountain area of the Morrison Quadrangle, Colorado.
SOURCE: Scott (1972).

their contrast to "background" conditions in the vicinity. Results are greatly dependent on the skill and experience of the interpreter (Soeters and van Westen, 1996).

Interpretation of aerial photographs, especially stereographic images, for identifying slope instability has long been accepted as a valuable landslide investigation technique. However, a number of new remote-sensing opportunities now exist for landslide investigators. Earlier satellite imagery with relatively low spatial resolution was of little use for landslide studies except for basic inventories of large regional extent. The comparatively recent advent of commercial satellites capable of providing images with 1-m, and even submeter resolution suggests that such satellite imagery will form an important component of future landslide studies.

BOX 3.4
APPROACHES TO LANDSLIDE MAPPING

The Inferential Approach

This approach is very common and relies on visual analysis of aerial photographs, Light Detection and Ranging (LIDAR) (see section 3.2) and other remote-sensing images, topographic and geologic maps, and field observations and historical data, to create interpretative maps of the extent and relatively activity of landslide features. Four major classes of maps may be produced by the inferential approach:

1. *Landslide inventory maps* show the spatial distribution of mass movements, represented either as affected areas to scale or as point symbols (Wieczorek, 1984).

2. *Landslide density maps* show landslide distributions by *landslide isopleths* (Wright et al., 1974).

3. *Landslide activity maps* are usually based on interpretation of aerial photographs taken at different times.

4. *Qualitative combination maps* result when a scientist uses individual expert knowledge to assign weights to a series of parameter maps and then sums these to produce a series of relatively homogeneous slope instability zones (Stevenson, 1977).

Limitations. Maps produced by the inferential approach, while rooted in direct observation, are strongly dependent on the experience and skill of the mapmaker. *Inventory, density,* and *activity* maps are costly to create and require repeated updating after major landslide-producing storms. *Qualitative combination maps* may be unreliable when insufficient field knowledge of the important factors prevents the proper establishment of factor weights, leading to unacceptable generalizations.

The Statistical Approach

The statistical approach consists of mapping a large number of parameters considered to potentially affect landslides, and subsequent (statistical) analysis of all potential contributing factors. This analysis hopefully identifies conditions leading to slope failures. The advent of digital elevation data has encouraged the use of two distinct statistical approaches:

1. *Bivariate statistical analysis* evaluates each factor map (e.g., slope, geology, land-use) in turn with the landslide distribution map, and weighting values based on landslide densities are calculated. Brabb et al. (1972) provided an early example of such an analysis. USGS personnel in Menlo Park, California later applied geographic information system (GIS) techniques to statistical landslide mapping (Newman et al., 1978; Brabb, 1984, 1987; Brabb et al., 1989). Subsequently several statistical methods have been applied to calculate the weighting values (van Westen, 1993).

2. *Multivariate statistical* models for landslide hazard zonation have been developed in Italy, mainly by Carrara (1983, 1988) and his colleagues (Carrara et al., 1991, 1992). All relevant factors are evaluated spatially within grid cells or by morphometric units. The statistical model is built up in a "training area," where the spatial distribution of landslides is well known. Then the model is extended to the entire study area, based on the assumption that the factors that cause slope failure in the target area are the same as those in the training area. Bernknopf et al. (1988) applied multiple regression analysis to a GIS data set, using presence or absence of landslides as the dependent variable and the factors used in a slope stability model (soil depth, soil strength, slope angle) as independent variables.

Limitations. Statistical approaches have the advantage of using an objective procedure for hazard delineation and are relatively inexpensive to create once the digital data are available and statistical analyses have been performed. Good results are found in homogeneous zones or areas with only a few types of slope instability processes. In more complex situations, very large data sets may be required because the methods do not make use of selective criteria based on professional experience. Another disadvantage of the statistical approach is that specific empirical relationships may be of limited generality; hence statistical relationships have to be determined for each study region, the boundaries of which are not clear.

The Process-Based Approach

This approach uses a deterministic or process-based analysis to delineate relative landslide potential. Quantitative theory for slope instability is applied using digital elevation data and other digital information, such as geologic attributes and vegetation cover. The slope instability theory is commonly coupled to process-based hydrologic models. This approach emerged in the past 10 years and is undergoing rapid evolution, driven in part by new observational technology (see below).

Despite problems related to collection of sufficient and reliable input data, deterministic models are increasingly used for hazard analysis over larger areas, especially with the aid of GIS techniques, which can handle the large number of calculations involved when determining safety factors over large areas. Yet this approach is applicable only when the geomorphic and geologic conditions are fairly homogeneous over the entire study area and the landslide types are simple

Limitations. The main problem with process-based methods is their high degree of simplification. Slope stability is often strongly dependent on local conditions, such as planes of weakness in bedrock, root strength, or groundwater conductivity, which are at present nearly impossible to map at the resolution needed over a large area. Hence, the controlling parameters in processed-based models can be difficult to estimate, and considerable uncertainty must be accepted in model results.

Perhaps the most important of the new remote-sensing tools is the use of airborne mounted lasers (LIDAR [Light Detection and Ranging]) to produce remarkably fine-scale topographic maps. Landslides occur where the landscape steepens, and small differences in topography can produce large differences in the likelihood of ground failure. Computer-based analysis of topography will increasingly play a crucial role in identification of the location and analysis of the relative potential for landsliding. High-quality digital elevation data must be one of the foundations of a national program for landslide hazard mapping and mitigation.

Currently, the best available digital elevation data derived from digitizing standard 1:24,000 USGS topographic maps (with 10-m spacing) are inadequate to perform this analysis. High-resolution topographic maps can now be made quickly over large areas using LIDAR methods, which allow a grid spacing of as little as 1 m. In forest terrain, LIDAR-based terrain maps can be more revealing than high-resolution aerial photographs (Figure 3.3). The first landslide hazard maps based on LIDAR imagery are now being reported and analyzed by various groups, including the USGS. These early maps show that high-resolution airborne LIDAR surveys can reveal previously unrecognized deep-seated landslides, can capture the sharp edges of small shallow landslide scarps and debris flow runout tracks, can show debris flow fans, and can greatly improve models for mapping shallow landslide potential. Much work lies ahead in learning how to exploit these data.

At present, airborne laser mapping is carried out by one federal group, two universities, and a variety of private companies. The National Aeronautics and Space Administration (NASA) Airborne Topographic Mapper (ATM) pioneered LIDAR surveying, with predecessors that go back to the 1970s. Two major contributions arising from the ATM flights have been the ALACE (Airborne LIDAR Assessment of Coastal Erosion) program and the mapping of glacier fields in Antarctica, the Arctic, and Greenland. Much of the coastline of the United States (excluding Alaska and Hawaii) has been covered by the ALACE program. The USGS and the National Oceanic and Atmospheric Administration (NOAA) have both established projects associated with the ALACE program. Unlike the coastal program, the USGS is using commercial sources to generate LIDAR for its work on fault and landslide hazards. The ATM has recently completed some flights to address inland problems such as floodplains development.

The USGS Coastal and Marine Geology Program participated in coastal erosion studies along the Washington, Oregon, and California coasts to assess the damage caused by the El Niño winter storms of 1997-1998 (USGS, 2003a). These studies used repeat LIDAR surveys to generate accurate topographic models of the coastal areas before and after the storms to strikingly illustrate topographic changes.

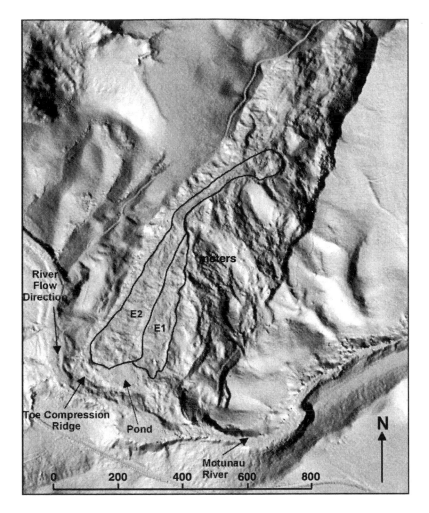

FIGURE 3.3 Computer-generated shaded relief map from LIDAR imagery showing the distinct topographic signature of a large deep-seated bedrock landslide complex 65 km north of Christchurch, New Zealand. Ground surface elevation points were recorded about every 2.6 m, with an estimated vertical error of 15 cm and horizontal error of 1 m. Small sinuous roads are visible on both sides of the landslide. McKean and Roering (2003) show that the data can be used to automatically detect and map the landslide complex. Two distinct earthflows are visible within the landslide; the one labeled E2 is currently active. Small compressional folds in the earthflows can be mapped accurately and automatically to provide insight into mechanical behavior and strength characteristics. Repeat surveys of such sites can be used to quantify additional landslide movement.

SOURCE: Reprinted from McKean and Roering (2003); copyright 2003, with permission from Elsevier.

The most active group in the USGS using LIDAR, located in Seattle, has collaborated with various county and state agencies to form the Puget Sound Lidar Consortium (http://duff.geology.washington.edu/data/raster/lidar/). Through grants from NASA, USGS, and various cities and counties, LIDAR surveys of a large area of the lowland southern Puget Sound have been completed. Another example of cooperation between state and federal agencies in the use of private company-derived LIDAR is the North Carolina Floodplain Mapping program (NCGS, 2003), which is supported through a cooperative agreement between the Federal Emergency Management Agency (FEMA) and the state. The state has established stringent surveying standards, tested the results of various commercial vendors, and disseminated the data. The entire State of Louisiana is also being surveyed by LIDAR to generate 1-foot contour maps, using a combination of state funds and support from FEMA. These data are easily available through the Internet (see LSU CADGIS Research Laboratory, 2003).

A number of states have completed or developed proposals for partial or complete statewide coverage by LIDAR mapping, and a coordinated effort to generate high-resolution topography nationwide would provide a consistent and uniformly high-quality data set for national and local applications. The two universities with LIDAR capability (University of Florida and University of Texas) are undertaking the much-needed role of training students and developing and testing the latest technology. Although they have both been involved in landslide studies, their primary focus is not on landslides, and their programs are oriented toward research. The National Science Foundation (NSF) has recently approved funding for an NSF-sponsored Center for Airborne Laser Mapping (NCALM), to be jointly managed by the University of Florida and the University of California at Berkeley. This center will provide research-grade LIDAR data and data analysis training to NSF-supported researchers and will be a valuable resource for advancing basic landslide research.

3.3 THE ROLE OF LANDSLIDE HAZARD ZONATION MAPS

There is currently no national inventory of known landslide and other ground failure hazards, although the USGS published *Landslide Overview Map of the Conterminous United States* in 1982—the only map with national coverage that delineates areas where large numbers of landslides have occurred or areas that are susceptible to future landslide events (Radbruch-Hall et al., 1982). Because the map is highly generalized, at a relatively small scale, and based on imprecise landslide information for much of the country, it is completely unsuitable for local planning purposes.

Several states or regions have reasonably detailed landslide inventories in map or database formats that, in many cases, have been compiled by state geological surveys in partnership with local governments or other state emergency organizations. The preparation of such maps was largely the result of mandates within the Stafford Act. A hazard mitigation clause is incorporated into FEMA-state agreements for disaster assistance, stipulating the identification of hazards and the evaluation of hazard mitigation opportunities as a requirement for federal assistance. In the case of state-declared disasters, some states also require the development of local hazard mitigation plans, and thus landslide zoning maps, as a prerequisite for receiving state emergency relief funds. Some areas have been subject to extensive landslide mapping and inventorying programs, in collaboration with federal agencies (e.g., Cincinnati, see Box 1.3; the San Francisco Bay Area, Box 3.5; and Colorado, Box 3.6).

BOX 3.5
LANDSLIDE HAZARD INVENTORY MAPPING
IN THE SAN FRANCISCO BAY AREA

Landslide hazards in the San Francisco Bay region received considerable attention in the early 1970s when local and state agencies, in collaboration with the USGS and the Department of Housing and Urban Development (HUD), undertook the San Francisco Bay Region Environment and Resources Planning Study to provide assistance for land-use planning. A large number of landslide inventory maps were produced (e.g., Radbruch-Hall and Wentworth, 1971; Taylor and Brabb, 1972), in some cases representing experimental methods of communicating landslide hazards to different user communities. In 1997, the USGS reviewed these earlier maps and created a six-part "landslide folio" containing digital data and map files outlining areas of potential landslide activity in the 10-county San Francisco Bay region (USGS, 1997). This folio includes digital maps showing the distribution of slides and earth flows (Wentworth et al., 1997), likely debris flow source areas (Ellen et al., 1997), and the extents of earlier detailed landslides inventory maps (Pike, 1997). Because the folio is directed explicitly toward the emergency services community, the information is presented at a content, accuracy, and scale commensurate with its needs. In addition, the folio could be used for identifying risks arising from slope failure in areas of future development. In anticipation that the folio might be used in determining landslide susceptibility for individual land parcels, the following disclaimer was included as a special note (USGS, 1997): *"Although we recognize the need for such assessments, it is neither possible nor appropriate to render site-specific judgments based only on the generalized information released in this report. A licensed geotechnical engineer or engineering geologist should always be consulted to evaluate any issues of slope failure for house lots and other properties."*

BOX 3.6
COLORADO LANDSLIDE INVENTORY MAPPING

A landslide inventory map for Colorado, along with several larger-scale regional and local landslide maps, was prepared by USGS geologists in the mid-1970s (Colton et al., 1976). In response to increased landslide activity throughout the state during several "wet years" in the mid-1980s, Colorado developed and published a statewide landslide hazard mitigation plan in 1988 (Jochim et al., 1988). The Colorado Geological Survey (CGS) and the Colorado Division of Disaster Emergency Services undertook this plan cooperatively, with the encouragement and financial support of FEMA and the USGS (Wold and Jochim, 1989; Rogers, 2003).

The 1988 Colorado landslide plan listed 49 locations where landslides posed serious risks to communities, areas, or facilities, divided into "landslide or rock fall" and "debris flow" categories. The landslide plan and its priority list incorporated all available information—it was designed to provide an action plan of manageable size that could be addressed with scarce staff and funding resources (Rogers, 2003). Landslide inventory maps and information, along with experience gained from local landslide hazard investigations, were crucial to development of the plan (Jochim et al., 1988).

Some 15 years later, the CGS has revised the priority list, and a new map of critical landslide areas has been developed (Rogers, 2003). This critical landslides map and revised list now include 46 locations. Several locations listed in 1988 have been dropped because their hazards have been substantially reduced, and other locations have been modified; some have been merged into a larger hazard area or corridor. The two categories of hazards—landslides or rock falls and debris flows—have been retained, but the locations within each category are classified into three tiers according to severity (Rogers, 2003):

- Tier 1 listings are serious cases requiring immediate or ongoing action or attention.
- Tier 2 listings are very significant but less severe, where adequate information exists, some mitigation is in place, or current development pressures are less extreme.
- Tier 3 listings have less severe consequences or primarily local impacts.

A national landslide inventory would form an important first step toward an appreciation of the true scope and distribution of landslide hazards. An accurate inventory would provide metrics for national policies and would greatly reduce the present uncertainty concerning the magnitude of economic loss and environmental damage caused by landslides. Individual components of a national inventory should be compiled by relevant state and local agencies in partnership with federal agencies. In an environment where choices must be made from a number of possible approaches, the proposed National Landslide Hazards Mitigation Program can play a vital role in evaluating methods, setting standards, and ratifying procedures. The USGS National Cooperative Geologic Mapping Program's STATEMAP component provides an excellent model for the provision of federal assistance to states, on a matching-funds basis, to carry out the mapping that ultimately will populate the national inventory. The combination of resources from both federal and state spheres, with input from local agencies where possible, offers the potential for effective identification and increased understanding of landslide hazards. The long-term management and maintenance of a national inventory would require commitments and resources that could best be provided by the federal government, through the USGS.

3.4 LANDSLIDE MONITORING TECHNIQUES

Monitoring existing landslides and sites of potential landslides plays an important role in landslide loss mitigation and landslide research. Monitoring serves several important purposes:

• Identification of initiation of sliding or increased rates of sliding to provide a basis for alarms and warnings that can reduce landslide hazards
• Determination of the depths and shapes of landslide masses as an adjunct to susceptibility and hazard mapping
• Development of improved understanding of landslide processes and triggering mechanisms
• Development of improved understanding of causative factors, such as earthquakes or high rainfall events
• Evaluation of the effectiveness of control measures.

Geophones, tape extensometers, piezometers, and rain gages have been used to monitor landslide movements for many years (USGS, 1999). However, monitoring technologies have improved rapidly in recent years, reducing the cost and expanding the range of monitoring possibilities. This has resulted in more widespread deployment of landslide warning systems that will also provide detailed data to enhance understanding of landslide mechanisms. Programmed, automated Electronic Distance Measurement systems have been employed in many locations where slopes must be monitored continuously. Global positioning system (GPS) and differential GPS (dGPS) technologies have been used to monitor landslide movements since the mid-1990s. The USGS conducted field tests at the Slumgullion landslide in Colorado (Box 3.7) and demonstrated that the observed rate of change of high precision GPS coordinates could be used to determine slide velocities within 10% on a time scale of a few days (Jackson et al., 1996). More recently, dGPS techniques have been used at the Panama Canal to measure landslide movements with extremely high accuracy. Satellite-based interferometric synthetic aperture radar (InSAR) technology is capable of measuring vertical ground displacements with

---

BOX 3.7
LANDSLIDE MOVEMENTS MEASURED BY GPS, SLUMGULLION LANDSLIDE, COLORADO

Movement of the Slumgullion landslide, in southwest Colorado, has been monitored using precise GPS measurements. In an initial monitoring effort (Jackson et al., 1996), station benchmarks consisting of 1-m sections of capped steel pipe were inserted flush with the landslide surface to provide the basis for relative and absolute movement determinations. The coordinates of station SEI1 (see below), which were assumed to be stable, were held fixed relative to the rest of the station network. Movements were calculated during a four-day measurement period and ranged from a few millimeters per day at the toe of the slide to 15 mm per day in the central part. More recent GPS-based monitoring efforts (Coe et al., 2000a, 2003) have taken advantage of improved precision from more advanced GPS technology to define the seasonal variation in slide movement, with the aim of understanding the influence of meteorological and climatic controls and their related surface and subsurface hydrological parameters. These measurements showed that velocities range up to 7.3 m per year, and that maximum velocities occur during the spring.

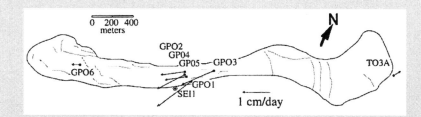

Location of GPS stations on the Slumgullion landslide during the initial GPS survey, with benchmark velocities calculated during a four-day measurement period. Dotted lines indicate the location of major structural elements of the landslide.
SOURCE: Jackson et al. (1996).

millimeter-level accuracy. Lower-cost, aircraft-deployed InSAR is being developed through NASA-sponsored research at Brigham Young University and is being used to monitor movement of the Slumgullion landslide.

Five landslides along U.S. Highway 50 in California have been monitored in real time using 58 instruments since heavy rains in January 1997 caused slope failures that destroyed three homes and blocked the highway (USGS, 1999). This network, operated in cooperation with the California Department of Transportation, provides engineers and geologists with early notification of landslide activity and with information useful in the design of remedial measures to halt these slides. Real-time data from one of these landslides are available to the public on the Internet (USGS, 2003b).

The USGS operates other remote real-time landslide monitoring sites. Near Seattle, Washington, real-time systems monitor the Woodway landslide and two other unstable bluffs that threaten a major railway (USGS, 2003c). Remote monitoring by the USGS in Fremont, California, monitors movement of a landslide that threatens homes, and in New Mexico, remote monitoring has recorded the effects of wildfire on slope stability (USGS, 1999). In Colorado, remote monitoring provides notification of ground movement caused by the very large landslide in DeBeque Canyon that threatens Interstate 70 (see section 7.3). In Rio Nido, California, the county government has assumed responsibility for operating a former USGS system that monitors a large landslide threatening more than 140 homes.

The national strategy proposal suggests that all partners (federal, state, local, private, and academic) be involved in (1) improving capabilities for monitoring, (2) monitoring landslides, and (3) establishing landslide warning systems. One part of the strategy proposes that real-time monitoring be integrated with NEXRAD (next-generation radar) data to improve warning capability. An existing partnership between USGS and NOAA's National Weather Service seeks to understand the relationship between rainfall intensity and duration with the thresholds for landslide initiation and the geologic determination of areas susceptible to landslides, so that real-time rainfall monitoring can be used for landslide hazard warning (e.g., Wieczorek et al., 2001, 2003). There is potential for broad application of such efforts to critical areas, with the likelihood that the involvement of FEMA in the partnership would assist emergency management. The USGS has taken the lead in applying the latest monitoring technologies in its volcano hazard program, and in its use of such technology for landslide monitoring at the Highway 50, Rio Nido, and Mission Peak landslides in California, it has worked with state agencies with the intention of ultimately transferring these capabilities to the state.

It is desirable that the USGS maintain a research program that stays abreast of rapidly advancing monitoring technologies, so that it is able to assist state and local government agencies to acquire the capability to deploy the latest systems where they can be used to enhance public safety. The USGS also has a logical role in using these technologies to develop detailed field data that will improve understanding of landslide mechanisms and in supporting university-based research toward this end. Private firms in the electronic and telecommunication fields will play an essential role in developing new technologies and bringing them to market. Private firms in the geotechnical consulting field will make use of these technologies to improve the state of practice in landslide monitoring for their private and public sector clients. The USGS, together with NSF and NASA, should also play a role in developing monitoring capabilities by supporting studies to explore new technologies and reduce their costs.

4

Landslide Loss and Risk Assessment

An understanding of the economic and societal impacts of landslides is essential for informed decisions that address the risks
from landslides and other ground failure hazards. Documentation of injuries and deaths, property damage, economic disruption, relief
and repair costs, and environmental consequences is part of such an
understanding. Undertaking risk assessments of prospective losses for
failure-prone areas is an allied and equally important process. Loss and
risk assessments are essential for

- establishing a sound rationale for risk reduction programs based
on documented economic and societal impacts;
- evaluating the cost-effectiveness of proposed interventions for
landslide-prone areas;
- creating mechanisms for risk sharing involving the public and
private sectors through insurance, special assessment districts, or other
financial risk pooling;
- partitioning responsibility for landslide-related cleanup, repair,
and rehabilitation costs; and
- understanding the noneconomic consequences of landslides events,
especially to the environment (e.g., damage to critical watersheds).

The terminology of loss and risk assessment can be confusing. The
term **loss assessment** is generally used in reference to *retrospective assessments* of the economic and societal consequences of a given event, and
more refined loss analyses go beyond an accounting of direct damages to

consider the economic and societal consequences of the event. The term **risk assessment** is generally used in reference to systematic *prospective analysis* of the extent of a hazard, the exposure of people and property to that hazard, the likelihood of a damaging event, and the likely resultant economic and societal consequences of that event. As articulated in the National Research Council "red book" (NRC, 1983) on risk assessments, risk assessments are the foundation for making decisions about the best means for managing a particular risk. Risk assessments can involve qualitative characterizations or more sophisticated quantitative calculations, and they can be based on scenarios describing individual events or probabilistic assessments across a series of potential events.

4.1 LOSS ASSESSMENT

The need for and problems in obtaining usable assessments of economic and other impacts of disasters constitute a problem that has been recognized in a number of recent studies. The basic problem is articulated in an NRC report addressing loss estimation for natural disasters: "There is no widely accepted framework or formula for estimating the losses of natural disasters to the nation. Nor is any group or government agency responsible for providing such an estimate" (NRC, 1999, p. vii). This issue, as it relates to landslides, was recognized with the 1980 publication of a U.S. Geological Survey (USGS) Circular that documented the costs of selected landslide events and called for systematic collection of loss information (Fleming and Taylor, 1980; also see Schuster, 1996; Schuster and Highland, 2001). Box 4.1 describes the conceptual and practical issues involved in undertaking systematic loss assessments.

The proposed National Landslide Hazard Mitigation Strategy identified the need for a ". . . framework for compiling and assessing a comprehensive data base of losses from landslides and other ground failure hazards, which will help guide research, mapping, and mitigation activities nationwide" (Spiker and Gori, 2000, p. 15). The Federal Emergency Management Agency and the insurance industry are identified as prospective leaders for two activities. The first is an assessment of the current status of data on losses from landslides and other ground failures nationwide. The second is to establish and implement a national strategy for compilation, maintenance, and evaluation of data on the economic and environmental impacts of landslides and other ground failures. The proposed strategy designates federal and state entities as responsible for creating a "robust national landslide hazards information clearinghouse system," local and private entities as responsible for collecting and distributing needed information, and the academic community as responsible for developing and sharing information.

**BOX 4.1
ISSUES IN UNDERTAKING *LOSS* ASSESSMENTS**

1. *Definition of losses:* Relevant considerations are the economic consequences, consisting of the value of harm to physical infrastructure and other direct impacts, together with the economic consequences of lost income, increased unemployment, and other indirect economic effects. These latter impacts are different, and typically much larger, than the dollar losses or damages directly associated with a particular event.

2. *Definition of direct and indirect components of loss:* The distinction between a direct and indirect loss is not always clear. Some consider lost revenues to be a direct cost, whereas others consider them an indirect cost. Loss accounting decisions can result in losses being double-counted or inappropriately counted.

3. *Value of life, injuries, and other noneconomic considerations:* Most loss estimates do not monetize these considerations but report them separately as other impacts.

4. *Meaningful losses or costs for different decisions:* Whereas economic losses are the relevant considerations for policy making and project evaluation, the relevant budgetary items for government reimbursement or insurance claims are the direct costs associated with damages, debris cleanup, and repair.

5. *Relevant level of geographic aggregation of costs:* As the geographic scope changes, issues arise concerning which losses, costs, and impacts accrue within a particular area and which accrue outside that area.

6. *Collecting damage and loss information:* Good data about losses are a key requirement. However, such data are difficult to obtain, because they have not been collected systematically in the aftermath of landslides (or any other natural disaster).

7. *Providing loss information in a timely and usable form:* Loss information must be provided in time for relevant decisions to be made and in a form that is meaningful for those decisions. At present, retrospective accounting of losses often requires months to collect even partial data.

Despite the identification in the national strategy proposal of other entities as appropriately leading loss assessment activities, the USGS and the Association of American State Geologists (AASG) have already taken the lead by establishing a partnership to undertake a loss assessment pilot project (Davis et al., 2003). The USGS provided funding to the AASG for a trial program to determine annual losses attributable to landslides in seven states (results from this pilot program were not available at the time of publication of this report). The committee commends the USGS and AASG for establishing this partnership and suggests that a series of such pilot projects will be necessary to determine optimum approaches to the collection and management of loss data that encompass both the economic and the social consequences of landslides. The committee endorses the USGS proposal for a "landslide loss information" clearinghouse to act as the focus for loss information (see section 5.3) and urges the USGS and state geological and other agencies to collaborate to ensure that appropriate protocols for data collection and storage are established as part of the National Spatial Data Infrastructure (OMB, 2002).

The committee further recommends creation of a Learning from Landslides (LFL) program to constitute a focal point for documenting the losses and other detrimental effects caused by landslides. Such a program could be modeled after the existing Learning from Earthquakes program funded by the National Science Foundation and coordinated by the Earthquake Engineering Research Institute. This LFL program would fund reconnaissance teams comprised of relevant specialists to examine and document notable landslide events and their impacts, including economic consequences (see section 6.3).

4.2 RISK ASSESSMENT

Risk assessments are the foundation for making decisions about the best means for managing a particular risk. The challenges of undertaking effective risk assessments include many of the issues associated with loss assessments, as well as others noted in Box 4.2. Figure 4.1 illustrates the role of risk assessment in guiding management of landslide risks, as presented in guidelines developed by the Australian Geomechanics Society (AGS, 2000).

Risk assessments provide informed options for risk management. As shown in the upper half of Figure 4.1, risk assessments are prospective analyses of the extent of a hazard, the exposure of people and property to that hazard, the likelihood of a damaging event, and the likely resultant economic and societal consequences of that event. Risk assessments can involve qualitative characterizations or more sophisticated quantitative calculations. They can be based on scenarios describing individual events or probabilistic assessments across a series of potential events. Although a number of American researchers and practitioners have been leaders in the development of landslide risk assessments (e.g., Einstein, 1988, 1997; Wu et al., 1996; Roberds et al., 1997), the range of procedures and their role in decision making are in general poorly understood in American practice, and consequently the use of formalized risk analyses is limited.

BOX 4.2
ISSUES IN UNDERTAKING *RISK* ASSESSMENTS

1. *Adequacy of understanding of process mechanisms and consequences:* Risk assessments entail projections of the consequences of future events for which a good understanding is required of the hazard itself, the exposure of people and property to that hazard, and the likely consequences of an event for that exposure.

2. *Adequacy of data for undertaking risk assessments:* Good data about the hazard and prospective losses are essential for undertaking risk assessments. Data and simulation capabilities are required to permit the prediction of events and their consequences on a localized basis.

3. *Predictions of likelihood of events and consequences:* Probabilistic risk assessment considers the likelihood of various events and the likelihood of various consequences following one or more of the possible events. This is a more refined approach than considering a particular scenario event.

4. *Consequences for life, injury, and other noneconomic considerations:* As with loss estimates, most risk analyses do not monetize these considerations. However, they are an important component that should be addressed.

5. *Relevant geographic focus for the risk assessment:* As with loss assessments, the consequences will differ depending on the geographic focus of the risk assessment.

6. *Addressing uncertainties:* Each step in undertaking a risk assessment is subject to uncertainty. Uncertainties result both from the probabilistic nature of landslide events and from a lack of knowledge of basic physics and damage fragilities associated with the events. The evaluation and communication of uncertainties are important issues.

7. *Providing risk assessments in a timely and usable form:* As with loss assessments, the information must be provided in time for relevant decisions to be made and in a form that is meaningful for those decisions.

A nationally coordinated approach to publicizing landslide risk assessments offers the opportunity for broader dissemination and understanding, particularly at the local level.

Risk assessments are not just technical undertakings. As emphasized in an NRC report that analyzed risk (NRC, 1996), risk assessments can be important processes for informing relevant stakeholders about potential consequences and for gaining consensus about appropriate steps to address

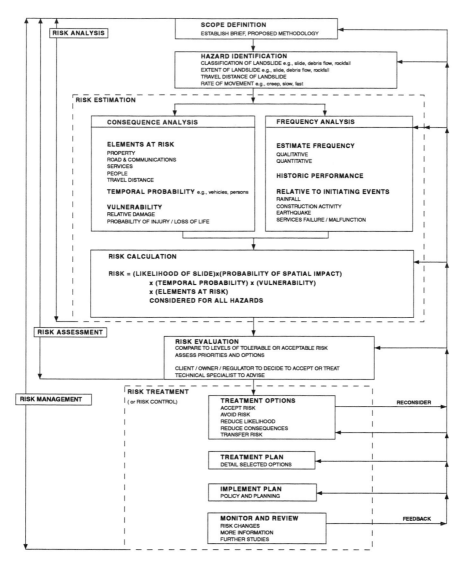

FIGURE 4.1 Schematic illustration of landslide risk assessment and risk management decision processes.
SOURCE: Australian Geomechanics Society (AGS, 2000).

potential harms. In this respect, an understanding of the risk posed by potential landslides is a central ingredient of determining appropriate risk management strategies to address that risk.

An example of qualitative approaches to risk assessment is the use of a scoring system for rock fall hazards by the Oregon Department of Transportation (Wu et al., 1996). Highway segments are evaluated for the likelihood of rock falls based on past frequency of rock falls, geological structure, and other considerations. The potential hazard is then considered along with potential for accidents (related to highway width and sight distance) to produce a rock hazard rating score. A number of other states have implemented similar approaches, and one role of a national strategy would be to ensure that all states have access to techniques and information that have been field-tested and refined.

An example of state-of-the-art quantitative risk assessment procedures for landslide problems is their use in Hong Kong (Box 4.3), where there is a substantial ongoing investment in improving hillside stability (Ho et al., 2000). Quantitative risk analysis has been applied to assess the cost of managing risk and the direct and indirect benefits that result, to optimize the allocations of available resources, and to identify areas of concern for improvement. A recent review of trial applications of quantitative risk assessment in Hong Kong (Lo, 2001) concluded that it can be a very valuable tool in landslide risk management.

Although the proposed actions described above for documenting landslide losses and risks are important and necessary for understanding the consequences of landslides, they should be perceived as one component of a dual approach. Risk assessments, together with loss analyses, are essential for informed decisions about the management of landslide risks. The committee strongly recommends that a national strategy for landslide loss reduction establish and promote the use of sound risk analysis methods for understanding landslide risks and making informed loss reduction choices. Because the state of the art of such methods is evolving, further development of landslide risk assessment methods and documentation of their use are important components of a landslide research program. Technical assistance in the conduct of landslide risk analyses should be central features of educational and other outreach activities established as part of the national program.

BOX 4.3
SAVING LIVES IN HONG KONG

On August 25, 1976, soon after 10:00 a.m., the fill slope immediately behind Block 9 of the Sau Mau Ping Estate in Hong Kong failed. The resulting mud avalanche buried the ground floor of the block killing 18 people. This followed equally horrific events in 1972 in which more than 100 fatalities occurred as a result of slope failures.

After an investigation, the government accepted the recommendation "that a control organization be established within the government to provide continuity throughout the whole process of investigations, design, construction, monitoring and maintenance of slopes in Hong Kong." Implementation of this recommendation has evolved into the internationally recognized Geotechnical Engineering Office. The basic mandate of the Geotechnical Engineering Office resides in enhancing public safety. This is recognized in government policy statements that describe targets for overall risk reduction. Current policy estimates are that by 2010, the overall landslide risk associated with man-made slopes will be less than 25% of the level that existed in 1977. The current annual budget for the Landslip Preventative Measures Programme to achieve this is approximately $US115 million per year (2003 estimates), expended primarily in upgrading existing high-risk slopes.

Analysis of known landslide fatalities in Hong Kong over the past 52 years (Chan, 2000) showed that, notwithstanding the extraordinary growth in Hong Kong since 1976, there has been a dramatic reduction in risk to public safety from landslide hazards. In addition, property values have been enhanced, although data on this aspect are not readily available.

The early phases of the slope management system in Hong Kong concentrated on inventory, mapping, and geological and geotechnical assessments. However, experience showed that enhancing public safety could not be achieved by technical considerations alone. Regulatory instruments, inspection and maintenance requirements, risk analysis, warning systems in response to intense rainfall, and an increased emphasis on community preparedness and response have all required development (e.g., Yim et al., 1999). The table presented below (from Malone, 1998) summarizes the components of the Hong Kong slope safety system. The Geotechnical Engineering Office has integrated all aspects of geotechnical engineering, together with the additional nontechnical tools required for effective risk management in a public setting.

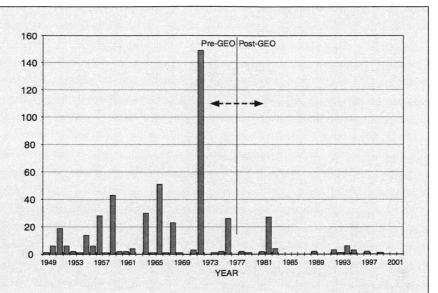

Known landslide fatalities in Hong Kong before and after establishment of the Geotechnical Engineering Office.
SOURCE: Modified after Chan (2000); figure courtesy of the Geotechnical Engineering Office of the Government of the Hong Kong Special Administrative Region.

Slope Safety System components	contribution by each component		
	to reduce landslip risk		to address
	hazard	vulnerability	public attitudes
policing			
cataloguing, safety screening and statutory repair orders for slopes	√		
checking new works	√	√	
maintenance audit	√		
inspecting squatter areas and recommending safety clearance		√	
input to land use planning	√	√	
safety standards and research	√	√	√
specialist works projects			
upgrading old Government slopes	√		
preventive works for old tunnels	√		
education and information			
maintenance campaign	√		√
personal precautions campaign		√	√
awareness programmes	√	√	√
information services	√	√	√
landslip warning and emergency services	√	√	√

Components of the Hong Kong Slope Safety System.
SOURCE: Malone (1998); reproduced with permission.

5

Loss Reduction Strategies

A successful loss reduction program must be based on effective application of landslide information at federal, state, and local levels. While recognizing that significant improvements in our ability to mitigate landslide losses undoubtedly will be developed in the future, a wide range of effective loss reduction measures exist now. Outreach programs and additional assistance measures are needed to help ensure that such loss reduction strategies are in fact used by state and local agencies and private entities.

The major responsibility for employing loss reduction measures for any particular project resides with the person or entity actually developing the project. All government units, including cities, special districts, counties, states, and federal agencies, have a responsibility to use appropriate loss reduction measures when undertaking public projects. Most development, however, is undertaken privately. In these cases, it is up to the government agencies that approve private development to ensure that appropriate loss reduction measures are followed. Professionals in the field share this responsibility to use the most up-to-date mitigation measures.

The National Landslide Hazards Mitigation Strategy proposal (Spiker and Gori, 2000) acknowledged that a successful strategy must include a mitigation component and that mitigation activities are generally undertaken by state and local governments, private businesses, and individuals. The proposal notes that a range of mitigation measures exist—including land-use planning and regulation, engineering, building codes, assessment districts, financial incentives and disincentives, emergency warning, and emergency preparedness—but also points out that there are impediments to the use of such measures. It then proposes the following actions:

1. Evaluate the impediments to effective planning and controls on development and identify approaches for removing these impediments.
2. Develop an education program for state and local elected and appointed officials to sensitize them to the risk and costs of landslide hazards.
3. Develop and disseminate prototype incentives and disincentives for encouraging landslide mitigation to government agencies, the private sector, and academia.
4. Evaluate engineering and construction approaches to mitigate landslide hazards and develop a national plan for research to improve these techniques.
5. Encourage implementation of successful landslide mitigation technologies.
6. Improve the ability to prepare for, respond to, and recover from landslide disasters.

5.1 IMPLEMENTATION OF LOSS REDUCTION MEASURES

The loss reduction measures suggested in the National Landslide Hazard Mitigation Strategy proposal (Spiker and Gori, 2000), presented above, form *part* of the range of items that should be considered in a comprehensive national loss reduction program. These measures are described only briefly in the strategy proposal. The education and public awareness components of loss reduction are dealt with in the following chapter; here the committee's comments and suggestions related to the other loss reduction measures are presented. Section 5.2 discusses a number of additional measures that should be considered as components of a national loss reduction strategy, and a commentary on the information collection, interpretation, dissemination, and archiving components of the strategy can be found in section 5.3.

Impediments to Effective Planning and Controls on Development. Although considerable information exists in some areas of the country with respect to landslides, in general, such information is not sufficiently known or used by local governments. In many cases, this is because negative information (e.g., information concerning natural hazards) is not highly attractive to local elected officials and the development community. Considerable literature exists in the social science field that addresses these issues—this literature must be assessed and the results employed in designing an effective outreach program. The Natural Hazards Center at the University of Colorado in Boulder[1] is an important resource in this respect, making available a substantial library on natural hazards and human responses to such risks. The center also hosts an annual workshop attended by a broad spectrum of federal, state, and local officials involved in natural disaster mitigation; academic researchers; and representatives of professional and nongovernmental organizations.

As noted earlier, most local governments do not have landslide hazard maps, and communities usually look to a higher level of government for mapping. In California, to comply with the Seismic Hazards Mapping Act, the state is preparing maps that identify areas subject to landslides and enhanced ground shaking. Prior to approval of a development project, cities and counties must require a geotechnical report on the subject property. In addition, property sellers or their agents must disclose information shown on these maps prior to property sale. Oregon has adopted similar legislation that specifically addresses fast-moving landslides (debris flows). This legislation directs the state to undertake mapping of landslide-prone areas and directs local governments to take actions to reduce potential damage from mapped landslides. Funding for this mapping program was provided by the Federal Emergency Management Agency's (FEMA's) Project Impact (FEMA, 2003). In addition to these examples from California and Oregon, some other state geological surveys have carried out extensive mapping of geological hazards (e.g., Alabama, Colorado [see Box 3.6], Kentucky).

As noted in Chapter 3, the substantial opportunities arising from high-resolution digital elevation mapping using LIDAR (Light Detection and Ranging) have prompted several states to undertake or propose partial or complete statewide LIDAR mapping programs. A nationally coordinated program to accelerate LIDAR mapping would provide significant assistance to local governments and make a major contribution toward landslide hazard mapping.

However, from a national perspective, state landslide hazard mapping activities are severely restricted by the minimal funds available. Some federal assistance has been provided (e.g., U.S. Geological Survey [USGS] demonstration mapping, FEMA Project Impact support), but there is a clear need for additional funding for this activity. The payoff from such state-level funding comes in the form of enhanced awareness and mitigation at the local level. Conditions vary across the country, but a balance between USGS, state, and local mapping should be established for each state. States could then implement legislation to require that local

---

[1]See http://www.colorado.edu/hazards/

governments take account of such information. In addition, guidelines could be developed and disseminated to assist states in their efforts to ensure that such maps are used effectively at the local level.

Encouraging Landslide Mitigation for Government Agencies, the Private Sector, and Academia. Emphasis should be placed on developing and disseminating incentives for mitigation. Communities must be convinced that it is in their own best interest to avoid the effects and repercussions of landslide disasters. Materials should describe such repercussions, which can include loss of life, injuries, financial losses to the private and public sectors, and lawsuits. Examples of successful land-use planning approaches should be documented and distributed. In some instances, land-use planning may lead to development of landslide-prone areas being avoided while at the same time allowing increased development in stable areas as compensation. In other instances, high land values may call for expensive engineering solutions for unstable land. Outreach activities, either as hard-copy publications or web-based information, are needed to inform communities of the seriousness of development in landslide-prone terrain and the advantages of recognizing and realistically dealing with the problem.

Another approach is to establish disincentives in the situation where development may be allowed in spite of hazards. For example, a disincentive can exist where a community allows a development in a landslide-prone area with the provision that the developer must disclose to property purchasers that they are buying property with a potential natural hazard. Otherwise, the developer may be disinclined to plan properly and the purchaser, particularly if notified at the last minute, is often disinclined to call off the deal. It is important that this information also be available to financial institutions (mortgage and insurance providers) that may be stakeholders.

Evaluation of Engineering and Construction Approaches to Mitigate Landslide Hazards and Development of a National Plan for Research to Improve These Techniques. Considerable information already exists that describes current engineering and construction approaches to mitigating landslide hazards. This information must be assembled and analyzed so that the major needs for the development of new engineering solutions to landslide problems can be identified. A specific research program could then be implemented to both improve existing techniques and develop new mitigation techniques. Analysis to determine if more effective and cost-efficient approaches can be developed would also be useful. In addition, new information continues to come from landslide experiences. This information should be analyzed to determine whether new engineering designs and practices are needed, and any such designs and practices should then be disseminated to the user community. In situations where legal liability issues might otherwise lead to a reluctance by industry to adopt new mitigation techniques, sponsorship and support of such techniques by government agencies may ensure more widespread application.

Implementation of Successful Landslide Mitigation Technologies. This can be accomplished largely through dissemination, via a national information clearinghouse (see section 5.3), of model approaches and good case histories. It is necessary to convince state and local agencies that although they might potentially be affected by landslides, there are effective ways to mitigate the hazard. This should be part of a coordinated outreach program undertaken as part of the national mitigation strategy.

Improving the Ability to Prepare for, Respond to, and Recover from Landslide Disasters. Landslide disasters normally affect small areas, compared to the more regional effects of events such as earthquakes, floods, or hurricanes. Nonetheless, they can cause major disruption in the affected and immediately surrounding areas. Accordingly, geologists, engineers, and emergency response professionals must be trained to understand the likely problems associated with a landslide, the emergency response that may be necessary, and the potential longer-term reconstruction needs. The full range of public utilities must be included in such preparation. Although such education and training should be included in the normal disaster plans and training programs at the local level, there is an important role for the national mitigation strategy to encourage such efforts and make available supporting materials.

5.2 ADDITIONAL LOSS REDUCTION MEASURES

The USGS proposal (Spiker and Gori, 2000) recognized and was responsive to many issues related to loss reduction activities; however, several additional areas should be addressed.

Standards of Care for Landslide Mapping and Engineering. Local regulations not only must require detailed landslide hazard mapping, but must ensure that the quality of hazard mapping meets appropriate standards of care. Such standards for hazard mapping and interpretation must be spelled out in local regulations, and the maps and reports prepared on behalf of a developer should be peer-reviewed by a qualified geologist on behalf of the jurisdiction. Finally, the actual grading and construction must be approved by both the preparer and the reviewer of these reports and plans. These standards of care are necessary to ensure that the final product meets the approved design. Institutionalization of such requirements in local regulations will help ensure that geological hazard mapping is used effectively. Model standards of care should be developed as a component of the national strategy to assist in the administration of local regulations, and an ultimate goal of partnership activity in this area would be incorporation of model standards of care into the Uniform Building Code and other building codes.

Linking Hazard Mapping to Land-Use Planning. Land-use planning approaches at the local level consist of general plans, zoning regulations, and subdivision regulations. Mapping, even at the standard USGS topographic map scale (1:24,000), can be useful to a community in helping shape its general plan for future development. Major areas of potential instability can be identified as needing special investigations or, in some instances, designated for open space use. In extreme conditions, general plans may propose cluster development in which construction is limited to stable areas and unstable areas serve as open space attendant to the development. The creative use of geological hazard information in preparing general plans, however, depends on a staff that has had training in the application of geologic information to plan making. Recognition that there are often negative implications from the identification of landslide-prone areas, in the form of decreased valuations, emphasizes the important support role that nationally accepted standards of practice have for staff at the local level.

Local zoning regulations stipulate, usually in great detail, how land can be used. Some communities have developed landslide matrices that are included in the zoning ordinance or as an adjunct to the ordinance. These matrices correlate categories of land stability with permitted or recommended land uses; they reflect a certain level of risk that the community has accepted. Two examples of this system occur in Morgan Hill and Portola Valley, both in California (Spangle Associates, 1988; Morgan Hill, 1994). A similar approach, although not incorporated in zoning, was prepared for the Portland, Oregon, regional government (DOGAMI, 1998). In some cases, not only are land uses identified, but they may be tied to general building types. Somewhat similar approaches have been applied in other parts of the country (e.g., Cincinnati, Ohio; see Box 1.3).

It is at the subdivision stage (neighborhood zonation maps, see Box 3.2) that the future pattern of ownership and land use is firmly established. It is at this stage that a community must demand the most detailed information about geologic hazards. The preparation of subdivision regulations that address geologic hazards is not complicated—a more difficult task is to ensure that the regulations are properly administered. In addition, elected officials have to be convinced of the need for such regulations, and there must be both an administrative process and a staff capable of implementing the regulations. Guidance and assistance, in the form of publications and effective outreach, should be provided to local governments as part of the national strategy outreach, so that they are better able to incorporate landslide mitigation provisions into their general plans, zoning regulations, and subdivision regulations.

Protection of Existing Development. Unfortunately, many urban areas were developed prior to the adoption of good hazard mapping and review practices in landslide-prone areas. Each year, the news media report cases in which houses are damaged due to differential settlement, lost in landslides, or in the worst cases, destroyed by debris flows. Although most of these cases result in financial losses, in the case of debris flows the results are often much more catastrophic and can include loss of life. For developed landslide-prone areas, the developers have left the scene in most cases and property owners face the prospect of potentially or actually losing their homes. The problems for governments when landslides impact government property are less serious than for homeowners, since the loss usually represents a small fraction of a government's assets. However, with private property the loss can represent most of the owner's assets. It is in this area that governmental assistance is needed.

Landslide Insurance. Landslide insurance could potentially provide a financial mechanism for spreading the costs of addressing or recovering from landslides among broad categories of those at risk, while also being linked to incentives for reducing risks. Costs arising from landslides are already a significant charge to insurance companies as those affected by landslides engage in litigation to determine fiscal responsibility for remediation costs. Landslide insurance is not presently offered because of the "adverse selection" of potential policyholders (Olshansky, 1996). That is, without mechanisms for expanding the pool of policyholders, only those who are at greatest risk would purchase policies, making it financially infeasible to offer insurance. In addition, landslide loss records are insufficient for establishing risk-based rates for landslide insurance. These problems are similar to those associated with establishing a viable program for earthquake insurance.

The National Flood Insurance Program (NFIP) has often been cited as a model that could be applied to landslide hazards, and in fact there have been instances in which the NFIP has covered damage caused by mudslides. In this program, the federal government produces maps of areas that are subject to flooding across the nation. In order for local governments to have properties in their jurisdiction eligible for the NFIP, they must adopt local ordinances to ensure that losses from flooding will be minimized. The same type of requirements for cities and counties would be needed for landslide insurance. Another potential insurance approach would be to include landslide hazards into an all-hazard residential risk insurance program, as is the case for homeowners under New Zealand's governmentally subsidized natural disaster insurance program (Box 5.1) which covers earthquakes, landslides, tsunamis, and other natural hazards (Earthquake Commission, 2003). It is important to recognize that loss-sharing mechanisms such as insurance are only a means for spreading the financial burden of landslides; they accomplish nothing themselves in terms of reducing risks. However, if mitigation measures are required as conditions for insurance, they can result in significant loss reduction. Accordingly, losses arising from landslides should be incorporated into any future national program of disaster insurance as long as such a program includes encouragement of or requirements for landslide mitigation measures.

Assessment Districts and Homeowner Associations. At least two additional types of financial arrangement—special assessment districts and homeowner associations—are worthy of consideration for addressing landslide hazards. Both can provide financing for remedial actions prior to or in the aftermath of landslides. Special assessment districts are political jurisdictions created by state legislation for the purpose of taxing residents of those districts in order to carry out designated functions. California, for example, has statutory provisions that enable creation of Geological Hazard Abatement Districts. Another vehicle for remedial action is through homeowner associations. The covenants and conditions that govern the association can provide for assessments to care for areas that fail due to landslides. This can be critically important because in many developments, the homeowner association actually assumes the responsibility for maintenance of major areas and facilities that are owned in common by the association.

5.3 INFORMATION COLLECTION, INTERPRETATION, DISSEMINATION, AND ARCHIVING

The proposed National Landslide Hazards Mitigation Strategy incorporates a plan for information collection, interpretation, dissemination, and archiving (Spiker and Gori, 2000). The objectives of this component of the strategy are the following:

- Evaluate and use state-of-the-art technologies and methods for the

BOX 5.1
NEW ZEALAND NATURAL DISASTER INSURANCE

The New Zealand Earthquake Commission is a government agency that provides natural disaster insurance to residential property owners (Earthquake Commission, 2003; Geonet, 2003). Despite its name, it actually provides insurance coverage for the full range of natural disasters—earthquakes, volcanic eruptions, hydrothermal activity, tsunamis, and natural landslides. Coverage is primarily for property loss or damage, but there is also limited coverage for land loss resulting from any of these hazards or from storms or floods.

The Earthquake Commission has a three-pronged approach to natural disaster management that seeks to achieve the following:

1. Facilitate research and education about matters relevant to natural disaster damage and its mitigation, noting that "for the purposes of managing risk it is necessary to have knowledge of New Zealand's natural hazards and our vulnerability to them, since risk is a function of hazard and vulnerability." (Earthquake Commission, 2003).

2. Provide mandatory insurance coverage, obtained automatically for a premium of 5 cents per $100 covered with any home or contents fire insurance policy. This insurance coverage was instituted in 1993, although precursors date back to 1945.

3. Provide assistance with postdisaster recovery and replacement. One component of this role is the existence of rapid-response teams of engineering geologists and geotechnical engineers that can be mobilized within 24 hours of a major event, to provide appropriate advice to maximize public safety and to collect reliable and consistent landslide information.

dissemination of technical information, research results, maps, and real-time warnings of potential landslide activity.

• Develop and implement a national strategy for the systematic collection, interpretation, archiving, and distribution of this information.

The proposed national strategy will collect a large quantity of information that must be interpreted correctly and translated into usable products, and then effectively disseminated to users. The information must also be archived in a manner that will permit ease of access by interpreters and users at all levels and also will ensure permanent future access. The types of information to be collected and archived will include:

Some of the 69 houses wrecked or displaced as a result of the 1979 Abbotsford landslide in New Zealand. This 1960s-era suburb was constructed despite an earlier recommendation that the unstable soils were unsuitable for housing.
SOURCE: Photo courtesy of New Zealand Earthquake Commission; reproduced with permission.

- digital information, including images, maps, and reports;
- nondigital information, including hard copies of maps and images;
- nondigital technical research, loss estimation, and implementation reports;
- information from real-time monitoring;
- weather information and hazard alerts; and
- manuals, videos, and other training materials.

The committee offers the following comments on specific items in the plan to assist with the development of detailed implementation strategies.

Information Collection. An extensive program of information collection is essential for the development of maps and other interpretative products for hazard mitigation. Much of the information will be aerial photographs or electronic imagery from which high-accuracy topographic, geologic, and landslide inventory maps can be compiled. The proposed program anticipates that many workers will be involved in information collection, including USGS scientists, personnel from other federal agencies, personnel from state geological surveys and agencies, university researchers, and private consultants. Effective coordination is required to ensure that the most important information is collected, archived, and made available using efficient and high-quality procedures. In all cases, digital geospatial data should have associated metadata in accordance with Federal Geographic Data Committee (FGDC) guidelines (Box 5.2), as a component of the National Spatial Data Infrastructure (NSDI).

Interpretation of Information. Information must be interpreted by trained scientists and engineers. At the national level, scientists and engineers with the USGS and other federal agencies will have lead responsibility for collection and interpretation of information. At the state level, scientists and engineers in state geological surveys, highway departments, emergency response units, and other agencies will take the lead in preparing interpretive products. Contracts with universities and private companies should be used to expand the resource pool of qualified data collectors and interpreters. The products at the state level, however, will be developed for use at the local level by counties, cities, transportation and utility districts, and so forth. In many instances, local agencies will have to adapt or develop interpretive products for their specific application. Local personnel must to be trained to ensure that these products are interpreted and applied correctly, as well as widely used for hazard mitigation (see Chapter 6 for additional discussion).

Dissemination of Landslide Hazard Information. Because implementation occurs primarily at the local level, dissemination of data and information to the local level is a key element of the program. To ensure that dissemination occurs and that those receiving the information are able to interpret and apply the products, effective cooperative partnership programs must be developed between federal, state, and local partners. To provide incentives for local users to participate, a program of local grants and cost-sharing is needed. At present, the Federal Highway Administration has programs for disseminating technical information and providing financial aid to states in applying that information, and the USGS National Landslide Information Center has an active program for distributing landslide information to the public, researchers, and planners and to local, state, and federal agencies. The national strategy should contain a pro-

BOX 5.2
FGDC AND THE NATIONAL SPATIAL DATA INFRASTRUCTURE

The Federal Geographic Data Committee (FGDC) is a federal inter-agency committee responsible for facilitating and coordinating the activities of the National Spatial Data Infrastructure (NSDI). The NSDI encompasses policies, standards, and procedures for organizations to cooperatively produce and share geographic data. The 19 federal agencies that make up the FGDC are developing the NSDI in cooperation with organizations from state, local, and tribal governments; the academic community; and the private sector. The NSDI is relevant to any agency that collects, produces, acquires, maintains, distributes, uses, or preserves analog or digital spatial data, including all geographic information system activities, that are financed directly or indirectly, in whole or in part, by federal funds.

Different federal agencies have lead responsibilities for the various spatial data themes (e.g., USGS is responsible for all geologic mapping information and related geoscience spatial data). Lead agencies are required to populate each data theme, principally by development of partnership programs with states, tribes, academia, the private sector, and other federal agencies, and also to facilitate the development and implementation of FGDC standards for each theme.

To build and support the NSDI, any agencies that collect, use, or dis-seminate geographic information and/or carry out related spatial data activities are required to do the following, both internally and through their activities involving partners, grants, and contracts:

• Develop a strategy for advancing geographic information and related spatial data activities appropriate to their mission.

• Collect, maintain, disseminate, and preserve spatial information such that the resulting data, information, or products can be shared readily with other federal agencies and non-federal users.

• Allocate agency resources to fulfill the responsibilities of effective spatial data collection, production, and stewardship.

• Use FGDC data standards, FGDC Content Standards for Digital Geospatial Metadata, and other appropriate standards; document spatial data with the relevant metadata; and make metadata available on-line through a registered NSDI-compatible clearinghouse node.

• Coordinate and work in partnership with federal, state, tribal, and local government agencies; academia; and the private sector to efficiently and cost-effectively collect, integrate, maintain, disseminate, and preserve spatial data, building upon local data wherever possible.

• Support emergency response activities requiring spatial data in accor-dance with provisions of the Stafford Act and other governing legislation.

• Search all sources, including the National Spatial Data Clearing-house, to determine if existing federal, state, local, or private data meet agency needs before expending funds for data collection.

gram of partnerships and cost-sharing between each of the federal agencies and their state counterparts, combined with effective education and training programs, to ensure that information is disseminated and applied at the user level.

Archiving of Information. Repositories will be required to archive the data and make it accessible to all potential users. As much information as possible should be stored in digital format and made accessible through web sites. There should be a central point of contact, probably most efficiently managed through expansion of the existing National Landslide Information Center, with links to distributed data centers. Each state could provide links to these distributed data centers on state government sites, with emphasis on material most useful for its personnel and local user communities. Federal and state departments of transportation could include links to the central and distributed repositories of landslide information on their web sites. In effect, the primary national source for data and information would be a web-based national information clearinghouse. Digital images, digital data files, interpreted maps, and many other useful materials will be stored in these web sites, which should be interlinked to allow ease of access by users at all levels of interest. In addition to the web sites, archival libraries will be needed at the national and state levels to store nondigital documents and historical images. The USGS could locate regional landslide data repositories at each of its regional libraries, and each state geological survey or other lead agency participating in the program should secure a special section in its state library for archiving landslide information. These collections of archived information should be made readily available to users at all levels.

6

Public Awareness, Education, and Capacity Building

The U.S. Geological Survey (USGS) national strategy proposal states that "before individuals and communities can reduce their risk from landslide hazards, they need to know the nature of the threat, its potential impact on them and their community, their options for reducing the risk or impact, and how to carry out specific mitigation measures. Achieving widespread public awareness of landslide hazards will enable communities and individuals to make informed decisions on where to live, where to purchase property, or locate a business. Local decision makers will know where to permit construction of residences, business, and critical facilities to reduce potential damage from landslide hazards" (Spiker and Gori, 2000, p. 19). The strategy indicates that a range of activities, tailored to local needs, will be needed to raise public awareness of landslide hazards and encourage landslide hazard preparedness and mitigation activities nationwide:

- Develop public awareness, training, and education programs involving land-use planning, design, landslide hazard curricula, landslide hazard safety programs, and community risk reduction.
- Evaluate the effectiveness of different methods, messages, and curricula in the context of local needs.
- Disseminate landslide hazard-related curricula and training modules to community organizations, universities, and professional societies and associations.

This component of the strategy recognizes that knowledge—about the risk and about various options for mitigating the risk—must form the fundamental basis for providing support to communities at risk from landslides. Providing the data and information that will lead to knowledgeable communities is a complex undertaking that can best be accomplished with a combination of federal-level coordination and resources and state-level interaction with decision makers, professionals, and the general public. Education and information dissemination activities should be carefully targeted, recognizing that these different groups have differing education and information requirements. Outreach materials may have to be prepared in multiple forms targeted toward the different audiences, and scientific information, maps, and monitoring data must be available in appropriate forms for use in emergency management, land-use, and public and private policy decisions.

Disasters that occur provide an excellent opportunity to educate the public about natural hazards and lessons one can learn, especially with respect to improving hazard mitigation. Although this information is important for decision makers, it is of critical importance for educating the general public. The public can, in turn, bring pressure to bear on elected officials to take appropriate mitigation measures.

The committee agrees that the brief statement in the strategy proposal provides a broad outline of public awareness and education requirements. The remainder of this chapter presents comments and suggestions related to capacity building for decision makers and professionals, together with a brief description of a program—Learning from Landslides—for gaining the maximum information from past landslides.

6.1 EDUCATION FOR DECISION MAKERS

Planning for the education component of a national landslide hazard reduction strategy must recognize that the reduction of landslide losses through land-use planning and application of building and grading codes for private lands is the function of local government and will be implemented by local government decision makers. All of the advice and information supplied to decision makers will be of little value if they are not convinced of the need to take a recommended course of action and guided in understanding and using the available information. Well-prepared landslide risk analyses that relate to the geographic area of a jurisdiction are an excellent way of pointing out to decision makers the seriousness of a landslide hazard and the consequences of not taking appropriate mitigation steps. Well-illustrated studies that describe the risk of financial and personal losses can also be an effective tool.

The strategy should ensure that information about the need for landslide hazard mitigation and the available tools to achieve mitigation are available at the local level in all areas of the country where landslides pose a significant threat to urban and urbanizing areas, and that decision makers are provided with natural hazard decision support systems that are appropriate for their particular jurisdictions. Education and training at national, state, and local levels will be required, designed and targeted to ensure that they reach those most responsible for hazard mitigation. This must be tied to major natural hazard mapping efforts undertaken at the federal and state levels. As well as provision of maps, it should also include interpretive materials and guidelines to assist with using the maps in the local planning and regulatory environment, recognizing that localities throughout the nation differ in both their regulatory authority and their approach to reducing losses from landslide hazards. It is likely that programs specifically developed for lower levels of governments, with local involvement (e.g., participation of state geologists), will be more readily received and widely implemented at the local level. In some areas, regional associations of government entities might also become a focus of such programs, and regional offices of federal agencies (e.g., Federal Emergency Management Agency [FEMA], Federal Highway Administration [FHWA], U.S. Army Corps of Engineers [USACE]) could play a significant role in coordination and funding efforts. In addition, centers should be identified in each state to which cities and counties can turn for additional information and assistance. Such outreach must be a continuing effort rather than a single occurrence.

Several agencies, such as FHWA and FEMA, operate regular training programs for state and local officials that might serve as models for developing training courses for the landslide hazard mitigation program. Another model might be the successful training program developed and operated by the Colorado Geological Survey, where specialists from the state survey meet regularly in various parts of the state with engineering and building practitioners and local government personnel, including elected officials, to discuss geologic issues of concern to local communities. During these sessions, the agenda includes education components and discussion of specific problems. Similar education activities involving interaction between state and local officials, to target aspects of landslide mitigation, should be developed for all landslide-prone areas.

In addition, examples of successful past programs exist. Two programs financed by FEMA in California, the Southern California Earthquake Preparedness Project and the Bay Area Regional Earthquake Preparedness Project, attest to the success of vigorous outreach programs. In each instance, boards representing the user constituency provided overall guidance. Staff then went out to local communities with materials and assistance, and remained available on a continuing basis. More recently, the HAZUS program and Project Impact, both developed and promoted by FEMA, have been major steps in providing information and assistance to local communities. In addition, local training sessions have been held in some areas, with information about geologic hazards being supplied and practical applications explained. In the mid 1990s, the USGS funded a series of such workshops in the San Francisco Bay area. These workshops were conducted by earth scientists and planners and addressed local government employees (including engineers, building inspectors, planners, and consultants). The objective was to improve the level of understanding and performance at the local level. In the late 1970s-1980s, the USGS and the Department of Housing and Urban Development sponsored the San Francisco Bay Region Environment and Resources Planning Study (see Box 3.5) (USGS-HUD, 1971; USGS, 1974; Kockelman, 1975). That study focused on developing geologic information and guidelines for using the information in planning and policy making. It was highly successful in the region and serves as a model for what might be accomplished in other areas. These efforts have depended on a combination of federal assistance and a receptive audience. If loss reduction with respect to landslides is to occur, these types of programs must be instituted and promoted vigorously in landslide-prone regions.

Another avenue for increased interaction among the various practitioners would be to implement a formal arrangement among federal, state, and local governments to loan or rotate employees. This effort could use the Intergovernmental Personnel Act (IPA) program that presently exists and would only require active encouragement by senior management. Such cross-fertilization would help to create a working, effective network; would aid effective communication; and would broaden the experience base of the employees involved.

It is clear that there is no single correct approach to developing a strategy for providing the required education and information to decision makers. Rather, it will take the varied efforts of many agencies and individuals to be successful. Some of these efforts will require a high level of coordination, whereas others will depend on the actions of individuals under differing circumstances.

6.2 ASSISTANCE FOR PROFESSIONALS

An extensive amount of research and practice has occurred in landslide science and engineering over the past 150 years. Consequently, a substantial amount of valuable information regarding landslides exists, encompassing fundamental science; hazard identification; assessment and mapping; techniques and methods for evaluation and analysis; and strategies, tools, and procedures for mitigation. This knowledge is available from a wide range of agencies and locations, including the published literature (e.g., Turner and Schuster, 1996); federal, state, and local agencies; private engineering and applied science firms; and developers, planners, and others. One of the most valuable aspects of a national landslide mitigation program would be the efficient dissemination of existing and future knowledge to engineers, scientists, planners, developers, and other professionals who deal with various landslide issues on a day-to-day basis.

At present, there are a number of agencies and institutions offering continuing education programs in landslide assessment, analysis, and mitigation for practicing professionals, including FEMA, FHWA, USACE, some academic institutions, and several state highway departments. Many academic institutions offer undergraduate and graduate courses in geoscience and engineering that focus on, or include, landslide mitigation topics. However, despite the large number of education programs and the involvement of several agencies offering valuable guidance, there is no coordinated approach to ensure that the required education and guidance are provided to the wide range of professionals and offices having interests and responsibilities in landslide hazard mitigation.

There are at least five fundamental areas in which coordinated educational programs for professionals should be established: (1) the fundamental science of landslide mechanisms; (2) landslide hazard and risk assessment and mapping; (3) geotechnical engineering evaluation and analysis of landslides; (4) mitigation of active and potentially active landslides; and (5) social issues in landslide hazard mitigation. Educational programs for each of these five areas are described briefly below:

1. Understanding the fundamentals of landslide mechanisms is a basic requirement before the other issues and challenges can be addressed. For the most part, the mechanisms of various landslide processes are reasonably well known (see Chapter 2). The problem in predicting landslides is a lack of understanding regarding the detailed site conditions and subprocesses that change in time and space. Education in the fundamental mechanics of slope failure processes would provide the necessary scientific background for the scientists, engineers, and planners at the local level that deal with landslide problems on a frequent basis.

2. Planners, scientists, and local engineers must be trained in the methods and concepts of landslide hazard and risk assessment and mapping (and even, in some cases, in the use of maps). Many city, county, and state offices have the data systems (geographic information systems and databases) with which to begin to develop landslide hazard and quantita-

tive risk evaluation and mapping programs. Several localities, such as Cincinnati, Ohio (see Box 1.3), have already developed an impressive capability for landslide hazard and risk evaluation and mapping. Lessons learned from the Cincinnati program and other successful efforts should be considered during the development of a standard recommended approach and the preparation of educational programs for local and state agencies.

3. One of the most important activities following the occurrence of a landslide in a critical area is the geotechnical evaluation and analysis of the feature. There appear to be as many methods of geotechnical investigations as there are practitioners. One reason for the broad range of investigational procedures is the uniqueness of each landslide. There are, however, fundamental data requirements for every type of landslide investigation that should be determined before mitigation activities are initiated. Several chapters in *Landslides, Investigation, and Mitigation* (Turner and Schuster, 1996) provide an excellent foundation for the development of a standard geotechnical investigation program for engineers and geoscientists. The educational program for geotechnical investigations should focus especially on the roles of the geologist and the engineer in terms of each person's responsibility, the criticality of dialogue between the two at all phases of the investigation, and opportunities for synergy.

4. Upon completion of the geotechnical investigation of active or potentially active landslides, mitigation activities commence. Mitigation may include decreasing driving forces, increasing resisting forces, avoidance, or some combination of all of these approaches. Many state and local agencies and private engineering firms have substantial experience in developing and implementing engineering methods for landslide mitigation. These agencies and firms have developed methods that work in their area or in the various areas in which they have been involved. A concerted effort should be made to catalogue these experiences and methods and to develop a comprehensive educational program for agencies and firms in all areas. This would ideally be undertaken by a federal-state-local-private partnership, probably with state geological surveys acting as the principal point of contact because of the geological variability among states.

5. The social issues in landslide hazard mitigation are often more difficult to resolve than the technical issues. Generally, the social issues revolve around the mitigation alternative of avoiding landslide hazard areas when possible. Use of the alternative of avoidance for landslide mitigation requires that a large number of issues be addressed, many of which are dear to persons in the area affected. Consequently, those with the responsibility for developing mitigation plans have to be aware of all of the challenges associated with the social issues of landslide mitigation. In

this respect, the social science-focused activities of the Natural Hazards Center at the University of Colorado, Boulder, represent a valuable resource.

Professional societies such as the American Society of Civil Engineers, the Association of Engineering Geologists, the American Geological Institute, the American Institute of Professional Geologists, and the American Planning Association (APA) serve as conduits of information from researchers to practitioners and from practitioners to researchers. The recently established two-year partnership between USGS and APA to provide training and technical support to local planners (Gori et al., 2003) is an example of this type of activity. Professional societies are generally the source of model codes, handbooks, and professional training for their membership, who in turn use the information to improve the state of knowledge of landslide loss reduction in the private and public sectors. These all can become valuable assets in the dissemination of information on landslide hazards.

There are at least four areas in which guidelines and specifications could be developed from existing information that would offer immediate benefit:

1. landslide hazard and qualitative risk assessment;
2. methods and procedures for landslide identification, evaluation, investigation, and analysis;
3. tools, materials, and conceptual designs for landslide mitigation; and
4. decision making by the public.

6.3 LEARNING FROM LANDSLIDES

Lessons learned though investigation and documentation of landslides can result in important insights that are invaluable for reducing losses from future events. To accomplish this, a Learning from Landslides (LFL) program should be created to carry out field investigations of significant landslides and to record and disseminate their findings. Such a program would generate new knowledge and should lead to changes in practice across many disciplines. This LFL program would provide a forum to observe and document landslide-generating mechanisms, as well as the effects of landslides on the natural and built environment and their resultant social, economic and policy implications. Findings from the LFL program, available on the web, would stimulate new research in each of the interrelated fields that landslides impact. The Earthquake Engineering Research Institute, a highly regarded national nonprofit association of earthquake engineering academics and practitioners, has a well-developed Learning from Earthquakes program that is in the process of being expanded to include additional hazards (e.g., floods), and this program could serve as a model for the LFL program.

The LFL program should be overseen by a broad, cross-disciplinary steering committee that would develop criteria for selecting the landslides to be investigated, determine the level of investigative response, and select members and define the responsibilities of the investigating team. The range of expertise of team members would encompass geological and geotechnical engineering, transportation and public utilities, emergency management and response, urban planning, and public policy. The range of factors that would influence the choice of particular landslides to be investigated would include the location relative to public works or population centers, risk of death or injuries, potential secondary or consequential impacts, and so forth. The LFL program could provide small grants during the one- to five-year period after a major landslide to document lessons learned from the recovery and reconstruction process.

7

A National Partnership Plan—
Roles, Responsibilities, and Coordination

Any national effort to address landslide risks must take into account the numerous federal, state, and local government entities that are already involved in addressing aspects of the problem. In addition, relevant stakeholders also include owners and managers of vulnerable transportation and utility networks, insurers and financial institutions that have a financial stake in property at risk, and researchers who contribute to understanding the problems and determining potential solutions. Consequently, responsibility for the problems posed by landslides and the solutions to those problems are widely shared among different levels of government and among different entities at each level. Recognition of this shared responsibility emphasizes the need for and role of partnerships in developing and implementing a national landslide mitigation strategy.

Stating that partnerships should be the foundation for carrying out a loss reduction strategy says little in itself. Partnerships can take a variety of forms. Too often partnerships constitute agreements on paper that have little practical effect. Partnerships are often established with a template that treats all partners in a similar fashion. Yet, partners often differ in their expertise, resources, and commitment. Moreover, each of these aspects rarely remains constant so that changes over time require adjustments to the partnership. Periodic reassessments to confirm that existing partnerships are still effective and useful are an essential element of a partnership strategy. In short, any partnership approach must recognize the differences between partners and also allow for changes in partnership relationships over time.

A key starting point for considering landslide partnerships is the recognition that for a national policy to be effective, it must shape not only federal actions but also those of state and local governments, and ultimately those of private landowners. This fundamental reality confronts a "shared governance" implementation dilemma that is common to the range of natural hazards policies (May and Williams, 1986). On the one hand, federal officials have a strong stake in promoting hazard mitigation. On the other hand, in the aggregate, it is ultimately the actions of state and local governments and of individuals owning property in hazardous areas that directly affect the success of loss reduction efforts. The dilemma arises because federal officials have little direct control over the effectiveness of such local efforts and because in many instances, given other priorities and competing pressures, state and local entities are often unwilling or unable to take the requisite actions to reduce prospective landslide losses. This reluctance has been documented in a number of studies of state and local government hazard mitigation planning (e.g., Berke and Beatley, 1992; Burby and May, 1998; May and Deyle, 1998).

7.1 PARTNERSHIP PRINCIPLES

Recognition of the diversity of entities and the need for shared governance of landslide programs leads to consideration of a set of principles for guiding the formation of partnerships. These principles are derived from studies of the shared governance of hazard mitigation programs (May and Williams, 1986; Paterson, 1998; Godschalk et al., 1999) and from evaluation of existing geoscience-related partnership programs (e.g., NRC, 1994, 2001).

The principles to be considered in the formation of landslide mitigation partnerships are the following:

1. *The need to allow for multiple partnership arrangements:* Given the diversity of entities at federal, state, and local levels, it is clear that multiple partnerships must be formed. A balance must be struck between fostering many relationships and relying on only a few well-established networks. The former may be hard to manage, whereas the latter may not open desired new avenues. Potential partnership networks include those among state geological organizations, local governments, research partners (including academic entities), and federal agencies.

2. *The need to embrace existing intra- and intergovernmental arrangements while allowing for development of partnerships with nongovernmental organizations:* Existing relationships should form the basis for development of vital partnership networks. These include partnerships involving multiple levels of government carried out under cooperative agreements. One such

example is the National Cooperative Geologic Mapping Program's partnership arrangements for coordinating mapping requirements that entail partnerships among federal and state entities (STATEMAP) and with academic institutions to fund mapping research (EDMAP). In addition, it is useful to foster nongovernmental partnerships with nonprofit organizations that can serve as forums for addressing aspects of the landslide problem. Examples of such forums include the National Association of State Floodplain Managers (a network of state flood management officials) and the Open GIS Consortium (a forum of private and public entities involved in developing protocols for geographic information systems).

3. *The need to share costs and responsibility:* A key tenet of partnership arrangements is the sharing of costs and responsibilities. This is essential for providing all partners with a sense of ownership and to minimize dependence on the financial resources of one partner. The specifics of sharing, of course, must be worked out in advance to be equitable, and they should be flexible enough to allow for changes in the circumstances of different partners over time.

4. *The need to minimize funding strings:* Discretion, within broad boundaries, in the use of funds or in approaches to achieving program objectives is desirable for promoting innovative solutions and approaches. Given the diversity of landslide problems, the differences in expertise among relevant organizations, and the differences in needs, such discretion would seem to be paramount.

5. *The need to tie partnerships to programmatic objectives:* Too often, partnerships languish during endless discussions of purpose, governance, and roles. While some level of discussion is often essential for building trust among partners, by definition partnerships that are all process achieve little. Accomplishments are more likely if partnerships are defined with respect to carefully described programmatic objectives. Partnership objectives should be defined in a manner to permit gauging progress with respect to key goals, while also allowing for innovation and necessary change.

7.2 RECOMMENDED PARTNERSHIPS

Because the responsibility for mitigating landslide risk is so widely distributed, it is imperative that a national strategy be based on a set of partnerships involving federal, state, local, and nongovernmental entities. The committee recommends creation or continuation of multiple partnership relationships for carrying out a national landslide strategy. These partnerships, their roles, and makeup are summarized in Table 7.1. Each of the entries shows the functions and focal point of the partnership. In most instances, partnerships will necessitate the involvement of multiple

TABLE 7.1 Recommended Partnerships

Partnership	Functions	Makeup
Federal-level partnership	• Leadership of national strategy • Coordination and funding of research • Coordination and funding of other partnerships • Resolution of landslide issues for federal lands and facilities	• Federal coordinating organization comprised of key federal agencies, led by U.S. Geological Survey
Federal-state risk assessment partnership	• Hazard identification and mapping • Promotion of risk analysis and mitigation practices for landslides affecting state land and properties	• Cooperative program with state geological entities • Cooperative program with state departments of transportation and other relevant state entities
State, local, and nongovernmental partnership	• Promotion of risk analysis and mitigation practices for landslides affecting local and private entities • Educational outreach to the general public and relevant professions	• "User group" partnership of local entities and nongovernmental stakeholders, potentially formed as a nongovernmental users group.
Research partnership	• Research on process mechanics, monitoring techniques, loss and risk assessment methods, and mapping techniques • Guidelines development and outreach activities	• Partnerships between universities and both governmental agencies and other stakeholders.
International partnership	• Sharing of research and practices for addressing landslide risks • Cooperative follow-up to major international landslide events	• Bilateral agreements between federal landslide coordinating council and relevant entities in other countries • Participation of U.S. professionals in international activities (e.g., International Consortium on Landslides, see Box 7.2; Joint Technical Committee on Landslides, see Box 7.3).

levels of government and of nongovernmental organizations. As such, the envisioned partnerships are not intended to be mutually exclusive. The formation and operation of each of the partnerships should be guided by the overarching partnership principles outlined above.

The specifics of these partnerships will have to be refined as part of an implementation plan for the national strategy for landslide loss reduction. With this caveat in mind, the committee envisions each partnership as functioning in the following manner:

1. *Federal-level partnership:* A national strategy should recognize the need for an interagency organizational structure to ensure leadership and coordination at the federal level. There are several potential models for such coordination, including the Federal Geographic Data Committee, charged with overseeing the National Spatial Data Infrastructure; the Interagency Committee of Seismic Safety in Construction, charged with oversight of Executive Orders concerning seismic safety of federal facilities; and the National Earthquake Hazard Reduction Program, responsible for coordination of federal earthquake risk reduction programs.

A federal-level landslide hazards reduction coordinating committee would consist of representatives from the major federal agencies that address landslide risks. An appropriate chair would be the Director of the U.S. Geological Survey (USGS) (or designee), with participation on the committee by an extensive group of agencies with responsibilities in this area (e.g., National Park Service [NPS], Bureau of Land Management [BLM], Federal Emergency Management Agency [FEMA], National Science Foundation [NSF], National Aeronautics and Space Administration [NASA], U.S. Army Corps of Engineers [USACE], U.S. Forest Service [USFS], Federal Highway Administration [FHWA], Federal Railway Administration [FRA]). As appropriate, representatives of states and major national organizations (Association of American State Geologists [AASG], Association of Engineering Geologists [AEG], American Institute of Professional Geologists, American Geological Institute [AGI], American Planning Association [APA], American Society of Civil Engineers [ASCE]) could be asked to serve on subcommittees of the federal-level coordinating entity.

The key functions of the federal coordinating partnership would be leadership of the national strategy, coordination and funding of research, coordination and funding of other partnerships, and addressing common issues concerning federal lands and properties including standards and guidelines for such facilities.

2. *Federal-state partnership:* The federal-state partnership is envisioned as the central mechanism for promoting hazard identification, mapping, and risk analyses at the state level. This partnership could be based on

cooperative agreements between the USGS and state geological agencies, between the FHWA and state highway departments, and between other relevant federal and state agencies. In keeping with the partnership principles articulated here, the cooperative programs would be undertaken on a matching funds basis. Existing organizations such as the AASG could form important components of this intergovernmental partnership.

3. *State, local, and nongovernmental partnership:* This partnership is envisioned as a broad-based coordinated effort among state governments, local governments, environmental and other nongovernmental organizations, relevant professional associations, and financial stakeholders. In essence, these constitute a "users group" for mapping and risk management products, as well as the key groups for implementing landslide risk loss reduction measures at the local level. In addition to undertaking specific mitigation measures (e.g., see Box 7.1), such groups are important for developing and carrying out the educational functions discussed in this report.

Local educational and loss reduction programs could be promoted through demonstration programs and cooperative federal agreements. However, it is important to consider the feasibility of forming a nongovernmental user group entity that would have membership open to these designated entities and would serve as a mechanism for sharing and promoting loss reduction practices. Examples of nongovernmental users groups are the Open GIS Consortium and the GeoData Alliance— consortia of government and nongovernmental organizations that have a stake in data formats and use (NRC, 2001).

4. *Research partnership:* Earlier parts of this report described critical research gaps encompassing aspects of landslide process mechanics, monitoring techniques, loss and risk assessment methods, and mapping techniques. Research on these topics is envisioned as continuing, in collaboration with state, local, and nongovernmental partners, with increased funding through intra- and extramural research programs of federal agencies (e.g., NSF, NASA, USGS). One option would be the establishment of a research center to provide a focus for these geoscience and geological engineering aspects of landslide mitigation. This would complement the present social science-focused activities of the Natural Hazards Center at the University of Colorado, Boulder. Whereas the Boulder center acts as a national and international clearinghouse to provide information on natural hazards and human adjustments to these risks, a natural science-focused center would act as a clearinghouse for technical and loss information and would perform important educational and outreach functions. Such a center could be distributed among a number of institutions, perhaps modeled on the Earthquake Engineering Research Centers at State Univer-

BOX 7.1
PARTNERSHIP CASE HISTORY—ROCK FALL MITIGATION

Serious rock fall hazards exist in Vail, Colorado, where ledges of resistant limestone and sandstone form cliffs above residential structures. After a severe rock fall in May 1983, the Colorado Geological Survey assisted the town in assessing the rock fall hazard. The town and some property owners in Vail Village formed a Geologic Hazard Abatement District (GHAD) and mitigated much of the hazard by constructing a ditch and berm on the slope above the residences.

The ditch and berm were designed with the aid of rock fall simulation software developed by university researchers with financial support from the Colorado Department of Transportation. Another very serious rock fall occurred within the GHAD in 1997, both above and to the west of the protective envelope provided by the ditch and berm. The ditch and berm stopped all rocks that fell toward it, as predicted, but rocks falling outside this protection zone caused substantial damage to other condominiums.

This case history demonstrates a successful partnership between state and local governments, academia, and local citizens groups to develop and install an effective mitigation system. It also shows the limitations of such arrangements—financial constraints resulted in the ditch and berm being constructed over too short a distance.

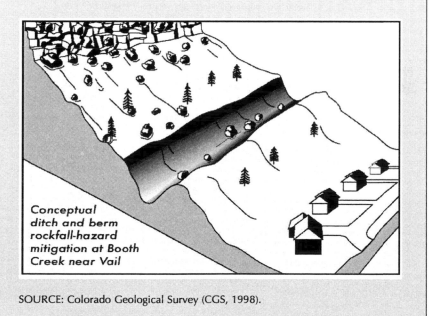

Conceptual ditch and berm rockfall-hazard mitigation at Booth Creek near Vail

SOURCE: Colorado Geological Survey (CGS, 1998).

sity of New York at Buffalo, University of Illinois at Urbana-Champaign, and University of California at Berkeley, with a charter to

- develop interdisciplinary curricula to provide the cross education for individuals working or desiring to work in landslide hazard mitigation;
   - conduct basic research into all aspects of the programmatic area; and
   - assist in the massive and key effort related to outreach; this would include outreach in its broadest sense—to scientists, engineers, planners, decision makers, and the public.

5. *International partnership:* Much can be learned from research and landslide experiences and practices in other countries, and the United States has an important role in sharing advances in this country with other countries. Consequently, international partnerships are an important component of a national strategy. The committee envisions the international partnerships as a set of bilateral agreements between the U.S. federal coordinating entity and relevant foreign partners. Participation of U.S. professionals in international activities could be managed under the auspices of the proposed natural science research center, including involvement in the International Consortium on Landslides (ICL; see Box 7.2) and the Joint Technical Committee on Landslides (JTC-1; see Box 7.3).

7.3 OVERVIEW OF FEDERAL, STATE, LOCAL, AND NONGOVERNMENTAL ROLES

The partnerships proposed here constitute the institutional frameworks for coordinating and carrying out a national landslide loss reduction strategy. Within these frameworks, federal, state, and local governments and nongovernmental entities have a range of roles and responsibilities. These are too numerous to identify in detail, but it is useful to consider the multiple levels of involvement of these entities and the key roles that are involved at each level (e.g., multisector partnership for the DeBeque Canyon landslide described in Box 7.4).

7.4 FEDERAL AGENCY ROLES

The federal role in a national landslide loss reduction strategy should be leadership, funding, and coordination of federal, state, local, and nongovernmental efforts. The primary mechanisms for carrying this out are envisioned as the partnerships identified in the preceding section. The national strategy proposal (Spiker and Gori, 2000) contains substantial information concerning the USGS role in the national strategy, but it has considerably less detail concerning the roles of other agencies. Although

---

BOX 7.2
INTERNATIONAL CONSORTIUM ON LANDSLIDES

The International Consortium on Landslides (ICL) is an international non-governmental and nonprofit scientific organization supported by the United Nations Educational, Scientific and Cultural Organization (UNESCO), the World Meteorological Organization, the Food and Agricultural Organization of the United Nations, the United Nations International Strategy for Disaster Reduction, and intergovernmental programs, such as the International Hydrological Programme of UNESCO, the Government of Japan, and other government bodies. The ICL was registered in August 2002 in Kyoto Prefecture, Japan, as a nonprofit organization, and the Secretariat is currently located in Kyoto.

The objectives of the ICL are the following:

• Promote landslide research for the benefit of society and the environment, and promote capacity building, including education, particularly in developing countries.

• Integrate geosciences and technology within the appropriate cultural and social contexts in order to evaluate landslide risk in urban, rural, and developing areas, including cultural and natural heritage sites; contribute to the protection of the natural environment and sites of high societal value.

• Combine and coordinate international expertise in landslide risk assessment and mitigation studies, thereby resulting in an effective international organization that will act as a partner in various international and national projects.

• Promote a global, multidisciplinary program on landslides.

The central activity of the ICL is the International Programme on Landslides, which sponsors research projects. Other activities planned include international coordination of landslide studies, exchange of information and dissemination of research activities, capacity building through meetings, dispatch of experts, development of landslide databases, and publication of the forthcoming ICL journal, *Landslides*.

---

the committee appreciates that the publication of one agency might avoid being particularly prescriptive concerning the roles of other agencies, a truly national strategy must nevertheless include a balanced description of the different roles. The committee's suggestions for the roles and activities of key federal agencies within the overall partnership framework—from an illustrative rather than comprehensive perspective—are as follows:

BOX 7.3
JOINT TECHNICAL COMMITTEE ON LANDSLIDES

The Joint Technical Committee on Landslides was formed in 2002 as a joint technical committee of the International Society for Soil Mechanics and Geotechnical Engineering (ISSMGE), the International Association for Engineering Geology and the Environment (IAEG), and the International Society for Rock Mechanics (ISRM). In representing these three international geotechnical professional societies, the committee is responsible for

1. discussing, advancing, and developing the science and engineering of landslides and constructed soil and rock slopes;
2. encouraging the collaboration of those who practice soil mechanics, rock mechanics, engineering geology, mining engineering, geomorphology, and geography, as applied to landslides on natural and engineered slopes;
3. fostering and organizing conferences, symposia, and workshops, including the international Symposia on Landslides, which are held at four-year intervals; and
4. contributing to the international congresses and conferences of ISSMGE, IAEG, and ISRM.

Department of the Interior. The Department of the Interior (DOI) is envisioned as serving as the chair of the federal agency coordinating council. Within DOI, important components of the strategy should be carried out by USGS, BLM, and NPS.

- *U.S. Geological Survey:* The USGS will have a central role in funding and carrying out research, in funding and carrying out cooperative hazard identification and mapping programs (in conjunction with state geological agencies), in serving as the central repository (clearinghouse) for geospatial data concerning landslide hazards, and in providing technical assistance and education. The USGS strategy proposal provides considerable information on the role that its Landslide Hazards Program, in partnership with state geological surveys, would play in a national mitigation strategy. The primary modification that the committee would suggest to the role proposed by the USGS is the addition of risk assessment, as an underlying principle guiding the prioritization of program activities, and the development and broad dissemination of landslide hazard risk assessment methods. Landslide Hazard Program activities should also include cooperative partnerships with other programs within the USGS, including

---

BOX 7.4
DEBEQUE CANYON LANDSLIDE: AN EXAMPLE OF A
SUCCESSFUL FEDERAL-STATE-ACADEMIC-PRIVATE SECTOR
PARTNERSHIP FOR LANDSLIDE MITIGATION

The DeBeque Canyon landslide is located on the south side of DeBeque Canyon in western Colorado. The landslide is very large—it extends about 1700 feet along the canyon and about 1,000 feet from the canyon top to the river. The landslide has more than 450 feet of elevation difference from head to toe (see aerial photograph below). Because of the importance of the combined I-70 and railroad transportation corridor, a major failure of the landslide has the potential to cause severe disruption to the Colorado and regional economy.

The landslide first affected Colorado transportation in 1924, when it disrupted the railroad. Later, U.S. Highway 6 was constructed through the canyon. Alignment improvements to this highway cut into the toe of the landslide in 1957, and subsequent movements affected the highway in 1958. Interstate 70 was constructed through the canyon in the early 1980s. Landslide investigations were conducted, and a design that incorporated additional loads on the slide toe was selected. The slide was stable until 1994 when slow creep failures began, and annual maintenance repairs have continued since then. More serious movement occurred in 1998, causing road damage and traffic delays on I-70.

Not only is the landslide huge; its movements are complex. The lower portions include a number of rotational elements, whereas the upper portions show translational movements. One very large block appears to be creeping slowly toward the canyon. Crevices have opened behind this block, which is composed of several extremely massive subblocks.

In response to the 1998 failures, the FHWA funded an emergency investigation program through the Colorado Department of Transportation. A partnership was formed to undertake the investigation, including the USGS, the Colorado Geological Survey, the Colorado School of Mines, and Golder Associates, a private consulting company. The Colorado Department of Transportation provided project oversight. The project team mapped the landslide and surrounding area, drilled test holes, designed an emergency response monitoring system, installed field instrumentation, modeled and interpreted the stability and potential failure modes of the landslide, and prepared cost-benefit mitigation scenarios. Each of the partnership members brought complementary skills to these tasks.

The DeBeque Canyon landslide continues to be monitored. The federal-state-academic-private sector partnership was successful in developing a cost-effective and timely response to a critical landslide hazard.

continued

BOX 7.4 Continued

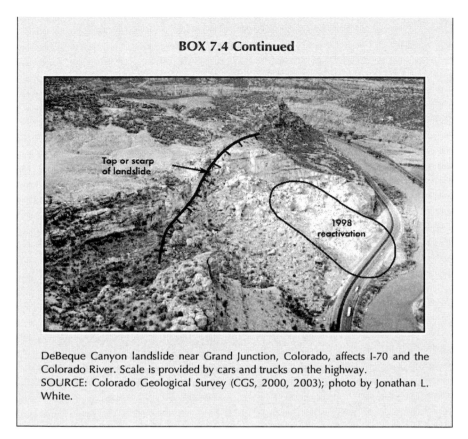

Top or scarp of landslide

1998 reactivation

DeBeque Canyon landslide near Grand Junction, Colorado, affects I-70 and the Colorado River. Scale is provided by cars and trucks on the highway.
SOURCE: Colorado Geological Survey (CGS, 2000, 2003); photo by Jonathan L. White.

the Coastal and Marine Program, to address submarine landslide hazards and with Biological Discipline programs for greater emphasis on identifying the ecological effects of landslides.

• *Other DOI Bureaus:* The BLM and NPS are essentially "customer" agencies, with USGS playing an important role in the provision of landslide hazard identification, mapping, and hazard mitigation on federal lands.

Department of Homeland Security—Federal Emergency Management Agency. Increasingly, it is clear that presentation of hazard information in terms of risks to people, risks to the built environment, and impacts on the social and economic spheres, can be an inducement to governments at all levels to take responsible action. The HAZUS program and Project Impact, both developed and promoted by FEMA, have been major steps in this direction, but they do not currently focus on landslides. FEMA has developed interactive exercises involving local communities in dealing with earthquakes, floods, and hurricanes. These exercises are designed to educate local government officials as to the types of problems they might face and pose questions concerning what steps could have been taken prior to the event in order to avoid or control the hazard. Such exercises could be developed for landslides.

At present, much of FEMA's landslide-related activity is focused on post-disaster cleanup activity following ground failure events. The committee's perception of FEMA's role as part of a national partnership is that it would place a much greater emphasis on landslide mitigation, including the provision of technical resources for the development of risk assessment methods, guidelines for risk assessments, and educational and other outreach materials concerning best practices for landslide loss reduction. FEMA's existing partnerships with state and local governments are important conduits for outreach programs. In addition, the recent post-landslide residential buyouts by FEMA using Hazard Mitigation Grant Program (HMGP) funds (see Box 7.5) show real promise for reduc-

---

BOX 7.5
FEMA-STATE BUYOUT PARTNERSHIPS

In 1998, the El Niño winter caused severe landslide damage to many parts of the western United States. In California, which was especially hard hit by the storms, FEMA and the Governor's Office of Emergency Services (OES) provided financial support under the FEMA HMGP for a statewide landslide acquisition or relocation program. Local communities that suffered El Niño landslides applied for buyout funds through OES, which reviewed and prioritized all proposals. OES then recommended eligible projects to FEMA for its review and funding. Although there was precedence for FEMA and California using HMGP funds to acquire properties that had been flooded repeatedly, this was the first time that money was provided for local governments to buy or relocate homes affected by landslides. Any structures that remained on the acquired properties were demolished, and the land was converted to open space, maintained by the appropriate local government entity. The total cost of the various approved acquisitions was $30 million, of which FEMA provided a 75% federal share ($22 million). Since that initial program in California in 1998, FEMA has undertaken similar property buyouts in other landslide-damaged areas in Oregon and Colorado.

ing future losses. To fulfill such an expanded role, FEMA would require additional funding, institutional commitment, and increased collaboration and interaction with other agencies.

Department of Defense—U.S. Army Corps of Engineers. As both an applied and a research organization, the USACE is envisioned as fulfilling a number of important roles within a national strategy, with particular leadership responsibility for engineered mitigation activities. The network of USACE districts and divisions will be important for technology transfer, to assist state and local government entities and other federal agencies with the planning and provision of engineering options to decrease landslide hazards. The USACE role should also include the development of guidelines and standards concerning engineered landslide mitigation.

Department of Commerce—National Oceanic and Atmospheric Administration (NOAA). Two of NOAA's organizations, the National Weather Service (NWS) and the National Ocean Service (NOS), should play important roles in a national strategy. A USGS-NWS partnership for landslide hazard warning, building on past partnerships, offers considerable potential for reducing the risk of injury or loss of life from rainfall-induced landslides. NOS and USGS partnered effectively to conduct coastal mapping activities for the Airborne LIDAR Assessment of Coastal Erosion (ALACE) project, and development of this partnership offers considerable potential for mapping coastal landslide hazards.

Department of Transportation—Federal Highway Administration and Federal Railway Administration. The FHWA is envisioned as having an important role in providing technical assistance and promoting loss reduction efforts undertaken at the state level by state highway agencies. The FRA has a similar role in providing technical assistance and promoting adoption of risk management practices for national railways. There is scope for both agencies to promote effective transfer of technology through demonstration projects.

Department of Energy. The national laboratories administered by the Department of Energy have a potential role as participants in research on landslide processes within their facilities addressing earth and environmental sciences.

Department of Agriculture—U.S. Forest Service. The USFS has an important role in addressing landslide hazards within federal forest lands. USFS geologists and engineers have extensive landslide expertise, particularly with the application of low-cost mitigation solutions along USFS access roads that are often located in unstable, mountainous regions. They have had a long tradition of conducting basic research on landslide mechanisms and the environmental consequences of landsliding associated with forest practices (e.g., Sidle et al., 1984). Increased landsliding mapping and research activity by the USGS on forested lands would contribute to the continued debate about how to minimize landslide occurrence.

National Aeronautics and Space Administration. NASA will have an important role in supporting basic research to develop and exploit remote-sensing techniques for hazard reduction. NASA will be a source of raw data on topography and other images (Synthetic Aperture Radar [SAR], hyperspectral, photographic, etc.) that can be "fused" to assist in landslide delineation, monitoring, and prediction. A recent report commissioned by NASA's Earth Science Enterprise (Solid Earth Science Working Group, 2003) explicitly calls for a "one time global mapping at 2- to 5-m resolution and 0.5-m vertical accuracy" of topography in the next 5 to 10 years. This would lead to a "continuously operating, targeted, high-resolution topographic mapping and change-detection capability" in the next 10 to 25 years (Solid Earth Science Working Group, 2003) that would contribute greatly to landslide hazard mitigation.

National Science Foundation. NSF is envisioned as having an important role in providing basic principal investigator-driven research support for understanding landslide hazards and processes. This would be conducted through a combination of existing extramural research programs and the proposed research center. The Interferometric SAR [InSAR] component of NSF's EarthScope initiative, requiring a dedicated L-band InSAR satellite to be developed and managed by a NASA-NSF partnership, offers considerable potential for landslide hazard studies. InSAR will be of particular value for the identification and quantification of movement in large landslides. The recent decision by NSF to support the Center for Airborne Laser Mapping will provide research-grade LIDAR data and training opportunities for researchers.

7.5 STATE AND LOCAL GOVERNMENT ROLES

The committee recognizes the paramount role of states, and particularly of localities, in carrying out a national landslide loss reduction strategy. States have important responsibilities in addressing landslide risks for state properties, in promoting local adoption of appropriate landslide loss reduction measures, and in identifying and mapping landslide hazards. These roles include involvement of the following:

• *State geological agencies:* State geological agencies, as part of the envisioned federal-state mapping partnerships, should continue and expand their roles in identification and mapping of landslide hazards. The discussion of education and outreach efforts (Chapter 6) emphasizes that for such maps to be effectively employed, state geological agencies must be proactive in providing technical assistance, education, and outreach programs for planners, geotechnical engineers, and others involved in risk analysis and loss reduction.

• *State highway departments:* State highway departments, as part of the envisioned federal-state highway landslide risk partnerships, should continue and expand their activities in addressing landslide hazards affecting state highways. These entities also have important roles in providing technical assistance to local public works departments.

• *State planning and building code agencies:* State planning and building agencies have important roles in setting and enforcing state standards concerning comprehensive planning requirements and building code provisions carried out by local governments. These entities also are involved in providing technical assistance to local governments. Appropriate consideration of landslide risks should be included in such state requirements.

• *State forestry departments.* State forestry and natural resource agencies have roles in addressing landslide potential and environmental impacts on state forest lands. In California, for example, the California Geological Survey provides landslide maps and field reviews of industrial forest lands to guide harvest plans.

• *Local planning and building departments.* Local planning entities (which might variously include county engineer offices, agricultural extension agents, soil conservation offices, etc.) have important roles in setting forth zoning and other planning provisions and in reviewing and approving plans for development and construction. As discussed in Chapter 5, land-use planning and development controls are essential instruments for addressing landslide risks. These entities also have important educational functions with respect to the general public and private sector planning and design professionals. Landslide risk should be part of these education and outreach efforts.

• *Local public works departments.* Local public works departments have responsibilities for municipal highways and infrastructure that is potentially at risk from landslides. These agencies should be aware of landslide risks and options for addressing them.

• *Emergency planning and response organizations.* Local emergency planning and response agencies have significant roles in promoting preparedness and in disaster response. As discussed in Chapter 5, improving the capabilities of these agencies to promote preparedness and to respond to landslide events is an important aspect of loss reduction.

7.6 ROLE OF NONGOVERNMENTAL ORGANIZATIONS

Owners and managers of vulnerable transportation and utility networks, insurers and financial institutions that have a financial stake in property at risk, engineering consultants, and university and other researchers who contribute to understanding the problems and potential solutions are relevant stakeholders. These entities are envisioned as being important players in the proposed local or nongovernmental user partnership and in the proposed research partnership. Their involvement is envisioned as consisting of sharing of technical expertise and experience, participation in partnership committees that develop approaches to promoting landslide loss reduction, participation in research activities, and financial contributions to user and research partnership programs. Key players among these entities include professional associations (e.g., AEG, APA, AGI, ASCE) and university researchers. Like other nongovernmental stakeholders, professional associations and university researchers are envisioned as being active participants in the user and research partnerships. In addition, the Public Risk Management Association, the Nonprofit Risk Management Center, and the Public Entity Risk Institute have collaborated to establish the Risk Management Resource Center to provide information on-line to help local governments, nonprofit organizations, and small businesses manage risks effectively (RMRC, 2003).

8

Funding Priorities for a National Program— Realizing the Vision

T his report provides a vision for a national strategy for mitigation of landslide hazards, expanding on the strategy outline presented in the U.S. Geological Survey (USGS) proposal (Spiker and Gori, 2000). Experience with landslides and advances in landslide research in the past decades have led to a better understanding of the physical processes of landslides, their potential consequences, and the means for reducing losses. Despite these advances, many gaps remain and attention to landslide risks in much of the country is haphazard at best. The vision of this report is one of a comprehensive national program that establishes a strong leadership role for the federal government but is based on partnerships with states, localities, and the research community, and emphasizes the translation of knowledge into practical applications.

8.1 FEDERAL FUNDING LEVELS

A comprehensive national program for addressing landslide risks requires a considerably increased level of federal funding for landslide partnership programs compared with current funding levels, and the committee believes that this ultimately will require more than the $20 million envisioned as the target budget for the USGS Landslide Hazards Program in the national strategy proposal (Spiker and Gori, 2000). The committee recognizes the reality that national budgetary considerations will determine the total annual funding provided to implement the strategy and emphasizes that it is the distribution of total available funding among the different components that is of paramount importance for an appropriately balanced national program.

This recommendation for substantially increased funding can be usefully compared with the current and recommended funding for National Earthquake Hazard Reduction Program (NEHRP). For this comparison, useful benchmarks are provided by a recent earthquake loss reduction research and action plan prepared by the Earthquake Engineering Research Institute (EERI) (EERI, 2003). Federal funding for the NEHRP program as of FY 2001 was approximately $100 million, distributed across the four partner agencies: USGS (48%), National Science Foundation (NSF) (30%), Federal Emergency Management Agency (FEMA) (20%), and National Institute of Standards and Technology (2%). The EERI research and action plan recommended annualized federal funding of $358 million for the first 5 years of a 20-year period (EERI, 2003). Annualized losses from earthquakes in the United States have been estimated as ranging from $4.1 billion (direct damage only) to $10 billion (including indirect losses), compared with estimated average annual losses from landslides of approximately $1 billion to $3 billion (NRC, 1985; Schuster and Highland, 2001). The landslide program funding level proposed here is equal to 20% of current earthquake funding and 5% of the EERI proposed funding level for a 20-year program of earthquake research and applications.

8.2 FUNDING PRIORITIES

The committee recognizes that the program proposed here requires a substantial funding increase that, for maximum efficacy, should be phased in over several years. In addition, it is important to recognize that the funding requirements of the program will change over its life. The national strategy proposal (Spiker and Gori, 2000) identified goals and implementation actions for a 10-year time scale, and accordingly, the following description of funding levels uses the perspective of a 10-year program. The proposed funding levels for such a program are based on three periods: (1) an *initial* funding phase of $20 million per year for three years; (2) an *established* program phase for three years of $35 million per year; and (3) a *mature* program period of $50 million per year for four years. As the program progresses from the initial, through established, to mature phases, the funding priorities also change from an initial emphasis on research, development of guidelines, and startup, to the later widespread implementation of landslide risk reduction measures through various partnership programs.

Table 8.1 provides a summary that compares the USGS proposed funding levels (Spiker and Gori, 2000) with the funding levels proposed

TABLE 8.1 Proposed Funding Levels for a National Landslide Hazards Mitigation Program (million dollars)

Program Element	USGS Proposed Annual Funding	Proposed Annual Funding		
		Initial (Years 1-3)	Established (Years 4-6)	Mature (Years 7-10)
Process mechanics	1.5	2	2	1
Monitoring techniques	2	3	2	1
Loss and risk assessment methods	0	2	3	1
Mapping techniques	2	3	3	1
Hazard identification and state mapping	12	4	10	15
Mitigation measures and programs	0	2	7	20
Learning from Landslides	0	1	2	2
Enhancing professional capabilities	0	1	3	4
Program management and staffing	2.5	2	3	5
Total	20	20	35	50

here for each of the three phases of a 10-year program. These figures are based on the committee's expert opinion regarding necessary funding levels for each of the program elements.

Initial Funding Phase (Years 1-3)—$20 million Annually. The emphasis during the initial funding phase would be on initiating a federal program, establishing the essential partnerships for carrying out the program, and developing the research and application foundation for the program.

Established Funding Phase (Years 4-6)—$35 million Annually. The emphasis once the federal program is established would be on moving the research and guidelines toward practical application through increased emphasis on cooperative mapping programs, demonstration programs concerning mitigation measures, and enhancement of professional capabilities.

Mature Funding Phase (Years 7-10)—$50 million Annually. The emphasis for the mature level of funding would be on sustaining the established initiatives with increased emphasis on translation of knowledge into practical applications involving risk management and mitigation programs that would be undertaken cooperatively with states and localities. Federal funds would be used for dissemination of guidelines and best practices, outreach efforts to encourage mitigation and loss reduction efforts at state and local levels, and cooperative mapping and demonstration programs. Federal funds are not anticipated as being used, other than as part of demonstration programs, to mitigate specific landslide risks.

8.3 FUNDING ALLOCATIONS

A different way of thinking about the recommended funding targets is to consider how the recommended federal funding might relate to the federal, state, local, and other partnerships that the committee envisions. Table 8.2 illustrates how the committee's proposed program targets could be allocated among different program partners across the three phases of a 10-year program.

The basic and applied research elements of the proposed strategy (process mechanics, monitoring, loss and risk assessment methods, and mapping techniques) are expected to be undertaken by a combination of intramural and extramural federally funded research. An important additional component of this research is the possibility of funding a landslide risk science and technology research center. Loss data information and

TABLE 8.2 Allocation of Federal Funds Among Partners (million dollars)

	Proposed Annual Funding		
Partnership Element	Initial (Years 1-3)	Established (Years 4-6)	Mature (Years 7-10)
Intramural and extramural federal research	7	7	2
Landslide research center	3	3	2
Loss data - Learning from Landslides	1	2	2
State mapping partnerships	4	10	15
Local and non-government mitigation partnerships	2	7	20
Other educational outreach	1	3	4
Federal program management	2	3	5
Total	20	35	50

consequences are anticipated as being collected as part of a Learning from Landslides program. Hazard and susceptibility mapping would be undertaken through the proposed cooperative federal-state mapping program. Hazard mitigation activities are envisioned as central to the proposed local and nongovernmental mitigation partnership. Educational outreach is envisioned as being undertaken through a variety of specialized grant programs.

9

Conclusions and Recommendations

Primarily because individual landslides usually affect limited local areas and individual landowners, damage resulting from landslide hazards has not generally been recognized as a problem of national importance and has not been addressed on a national basis. The absence of a coordinated, national approach to mitigating the detrimental effects of landslides has resulted in a reduced ability of state and local government agencies to apply the important lessons learned, often at considerable expense, in other parts of the country. As a result of a congressional directive, the U.S. Geological Survey (USGS) addressed the need for a national approach by preparing the National Landslide Hazards Mitigation Strategy (Spiker and Gori, 2000). This proposal describes in broad overview the nine major components, ranging from basic research activities to improved public policy measures and enhanced mitigation, considered essential to address hazards arising from landslides at the national level. **The committee agrees that a national approach to the mitigation of landslide hazards is needed and considers that the nine components briefly described in the USGS proposal are the essential elements of a national landslide hazard mitigation strategy.**

Responsibility for the problems posed by landslides, and for the solutions to those problems, is widely shared among different levels of government and among different stakeholders at each level. This shared responsibility emphasizes the role of the partnerships that will be required to develop and implement a national landslide hazards mitigation strategy. A key starting point for considering landslide partnerships is the recognition that for a national policy to be effective, it must shape not only federal actions but also those of state and local governments and ultimately those of private landowners. **The committee agrees that a national landslide hazards mitigation strategy should be based on partnerships involving federal, state, local, and nongovernmental entities.** The committee has defined five focal points for partnerships that will inevitably entail relationships within and among multiple levels of government and with nongovernmental entities:

1. partnerships between the federal agencies involved in landslide mitigation to provide leadership and national coordination;
2. partnerships between federal agencies and their state counterparts to promote hazard mapping and risk analysis at the state level;
3. partnerships between state agencies and local governments, nongovernmental groups, and private citizens to ensure that education and assistance is provided to the "front line" of mitigation activities;
4. research partnerships between federal agencies and academic institutions, in collaboration with state, local, and nongovernmental partners, to conduct research on landslide process mechanics, monitoring techniques, loss and risk assessment methods, and mapping techniques, and
5. international partnerships for global exchange of knowledge and techniques.

The description of the components of a national landslide hazards mitigation strategy in the USGS proposal is brief. The committee concludes that a more complete discussion of the comparative importance of each element of the proposed national strategy is required and a sense of priorities must be presented. The recommendations presented in the following paragraphs are designed to convey the committee's priorities for a national program:

The committee recommends that a national strategy for landslide loss reduction promote the use of risk analysis techniques to guide loss reduction efforts at the state and local levels. Because the state of the art of landslide risk analysis is evolving, further development of risk analysis methods, and documentation and dissemination of their use, are important components of the research and application program for a national landslide strategy. Use of risk analysis for guiding appropriate choice of landslide loss reduction tools should be an important element of the technical assistance and outreach programs provided to state, local, and nongovernmental entities. Development of guidelines and standards concerning best practices and promotion of those practices at state and local levels of government are important aspects of the proposed federal strategy.

The National Landslide Hazards Mitigation Program must play a vital role in evaluating methods, setting standards, and advancing procedures and guidelines for landslide hazard maps and assessments. National landslide information gathering and mapping should be undertaken within the proposed partnerships. The program must establish appropriate standards and procedures for the collection, long-term management, and maintenance of this information. Metadata must be associated with all data collected under the auspices of the program, in accordance with National Spatial Data Infrastructure protocols. Hazard zonation mapping must be developed for multiple mapping scales by utilizing best available technologies. Accurate terrain information is essential, and the landslide hazard mapping program must be based on the highest-resolution topographic data.

In order to provide tools for landslide hazard mitigation, it will be necessary to conduct basic research on monitoring techniques and on aspects of landslide process mechanics. An integrated research program is recommended in which intensive field studies are used to (1) improve site and laboratory characterization techniques; (2) develop new field monitoring methods; (3) obtain greater understanding of failure and movement mechanisms; and (4) develop and test models to predict failure timing, location, and ultimate mass displacement. Studies of debris flows, bedrock slides, and submarine landslides deserve greatest attention. Innovative remote-sensing technologies are now offering researchers the possibility of rapid and detailed detection and monitoring of landslides. Additional support to exploit these new technologies and develop practical tools for a broad user community is needed.

Improved education and awareness of landslide hazards and mitigation options, for decision makers, professionals, and the general public, must be primary components of a national landslide hazard mitigation program. Collecting and disseminating information about landslide hazards to federal, state, and local government agencies and nongovernmental organizations, planners, policy makers, and private citizens in a form useful for planning and decision making is critically important to an effective mitigation program. Such education and awareness efforts will be most effective if implemented at the outset of the program. If the national landslide hazard mitigation program is to materialize, broad-based acceptance, participation, and support are essential to its success.

The committee agrees that substantially increased funding will be necessary to implement a national landslide hazards mitigation pro-

gram. The committee considers that the figure of $20 million, presented in the USGS proposal as the amount required to support an enlarged Landslides Hazards Program within USGS, would provide an adequate basis for the initial stages of a national strategy with a 10-year target for achieving substantial loss reduction goals. However, the committee considers that the distribution of funding should progress from an initial emphasis on research, development of guidelines, and startup to the later widespread implementation of landslide risk reduction measures through various partnership programs. The committee considers that additional increases—to annual funding of $35 million for years 4-6 and $50 million for years 7-10 and beyond—will be required to support these later parts of the program. The committee recognizes the reality that national budgetary considerations will determine the total annual funding provided to implement this strategy and emphasizes that the distribution of total available funding among the program's different components is of paramount importance for an appropriately balanced national program.

The committee commends the USGS for undertaking the important initial steps toward a comprehensive national landslide hazards mitigation strategy. The committee recommends that the USGS—in close partnership with other relevant agencies—produce the implementation and management plans that will provide the practical basis for an effective national strategy that can be applied at the local level.

References

Anderson, L.R., J.R. Keaton, T. Saarinen, and W.G. Wells II, 1984. *The Utah Landslides, Debris Flows, and Floods of May-June 1983*. National Research Council, Washington, D.C., 96 pp.

AGS (Australian Geomechanics Society), 2000. Landslide Risk Management Concepts and Guidelines. *Australian Geomechanics*, 35(1): 49-92; reprinted in *Australian Geomechanics*, 37(2), May 2002.

Bardet, J.P., C.E. Synolakis, H.L. Davies, F. Imamura, and E.A. Okal, 2003. Landslide Tsunamis: Recent Findings and Research Directions. *Pure and Applied Geophysics*, 160(10-11): 1793-1809.

Berke, P.R., and T. Beatley, 1992. *Planning for Earthquakes: Risk, Politics, and Policy*. Baltimore: Johns Hopkins University Press.

Bernknopf, R.L., R.H. Campbell, D.S. Brookshire, and C.D. Shapiro, 1988. A Probabilistic Approach to Landslide Hazard Mapping in Cincinnati, Ohio, with Applications for Economic Evaluation. *Bulletin Association of Engineering Geologists*, 25(1): 39-56.

Bernton, H., 2000. Mount St. Helens' debris still sullies Toutle River. *Oregonian*, January 16, p. D1.

Brabb, E.E., 1984. Innovative Approaches to Landslide Hazard and Risk Mapping. Pp. 307-324 *in Proceedings, 4th International Symposium on Landslides*, Toronto, Canada, Canadian Geotechnical Society, 1.

Brabb, E.E., 1987. Analyzing and portraying geologic and cartographic information for landuse planning, emergency response and decision making in San Mateo County, California. Pp. 362-374 *in Proceedings, GIS'87*. American Society of Photogrammetry and Remote Sensing, Falls Church, VA.

Brabb, E.E., E.H. Pampeyan, and M.G. Bonilla, 1972. *Landslide Susceptibility in San Mateo County, California*. U.S. Geological Survey Miscellaneous Field Studies Map, MF360, scale 1:62,500.

Brabb, E.E., F. Guzzetti, R. Mark, and R.W. Simpson, 1989. The Extent of Landsliding in Northern New Mexico and Similar Semi-Arid Regions. Pp. 163-173 *in* D.M. Sadler and P.M. Morton (Eds.), *Landslides in a Semi-arid Environment*. Publications of the Inland Geological Society, 2.

Burby, R.J., and P.J. May, 1998. Intergovernmental Environmental Planning: Addressing the Commitment Conundrum, *Journal of Environmental Planning and Management* 41(1): 95-110.

Carrara, A., 1983. Multivariate Models for Landslide Hazard Evaluation. *Mathematical Geology*, 15(3): 403-427.

Carrara, A., 1988. Landslide Hazard Mapping by Statistical Methods. A "Black Box" Approach. In: *Workshop on Natural Disasters in European Mediterranean Countries*, 205-224.

Carrara, A., M. Cardinali, R. Detti, F. Guzzetti, V. Pasqui, and P. Reichenbach, 1991. GIS Techniques and Statistical Models in Evaluating Landslide Hazard. *Earth Surface Processes and Landforms*, 16(5): 427-445.

Carrara, A., M. Cardinali, and F. Guzzetti, 1992. Uncertainty in Assessing Landslide Hazard and Risk. *ITC-Journal 1992-2*: 172-183.

CGS (Colorado Geological Survey), 1998. Colorado's Geologic Hazards—How to Live with Them: Unstable Slopes. *Rock Talk*, 1(2): 4-6.

CGS (Colorado Geological Survey), 2000. Who Cares About the Colorado Geological Survey? *Rock Talk*, 3(1): 1-2.

CGS (Colorado Geological Survey), 2003. The DeBeque Canyon Landslide. Online; available at http://geosurvey.state.co.us/pubs/debeque/debeque.htm; accessed October 2003.

Chan, R.K.S., 2000. Hong Kong slope safety management system. Pp. 1-16 *in Proceedings of the Symposium on Slope Hazards and Their Prevention.* Jockey Club Research and Information Centre for Landslip Prevention and Land Development, University of Hong Kong.

Coe, J.A., J.W. Godt, W.L. Ellis, W.Z. Savage, J.E. Savage, P.S. Powers, D.J. Varnes, and P. Tachker, 2000a. *Seasonal Movement of the Slumgullion Landslide as Determined from GPS Observations, July 1998-July 1999.* U.S. Geological Survey Open-File Report 00-101.

Coe, J.A., J.A. Michael, R.A. Crovelli, and W.Z. Savage, 2000b. *Preliminary Map Showing Landslide Densities, Mean Recurrence Intervals, and Exceedance Probabilities as Determined from Historic Records, Seattle, Washington.* U.S. Geological Survey Open-File Report 00-303.

Coe, J.A., W.L. Ellis, J.W. Godt, W.Z. Savage, J.E. Savage, J.A. Michael Jr., J.D. Kibler, P.S. Powers, D.J. Lidke, and S. Debray, 2003. Seasonal Movement of the Slumgullion Landslide Determined from Global Positioning System Surveys and Field Instrumentation, July 1998-March 2002. *Engineering Geology*, 68: 67-101.

Colton, R.B., J.A. Holligan, L.W. Anderson, and P.E. Patterson, 1976. *Preliminary Map of Landslide Deposits in Colorado,* U.S. Geological Survey Miscellaneous Investigations Series Map I-964.

Cruden, D.M., and D.J. Varnes, 1996. Chapter 3: Landslide types and processes. Pp. 36-75 *in* A.K. Turner and R.L. Schuster (Eds.), *Landslides: Investigation and Mitigation.* Special Report 247, Transportation Research Board, National Research Council, Washington, D.C.: National Academy Press.

Davis, J.F., T. Berg, R. Schuster, L. Highland, P. Bobrowsky, and P. Gori, 2003. Consensus Strategies to Acquire Accurate Landslide Damage Cost Data and Apply Insights to Mitigation. Annual Meeting, Geological Society of America, Abstract 2003.

DOGAMI (Oregon Department of Geology and Mineral Industries), 1998. *Using Earthquake Hazard Maps. A Guide for Local Governments in the Portland Metropolitan Region.* Open-File Report O-98-04, Portland, Oregon.

Earthquake Commission, 2003. *A New Zealand government agency providing natural disaster insurance to residential property owners.* Online; available at http://www.eqc.govt.nz/; accessed June, 2003.

EERI (Earthquake Engineering Research Institute), 2003. *Securing society against catastrophic earthquake losses, a research and outreach plan in earthquake engineering.* Oakland, Calif. EERI, April 2002. Online; available at http://www.eeri.org/; accessed June 2003.

Einstein, H.H, 1988. Special lecture: landslide risk assessment procedure. Pp. 1075-1090 in C. Bonnard (Ed.), *Proceeding, 5th International Symposium on Landslides, Lausanne, Vol. 2.* Rotterdam: A.A. Balkema.

Einstein, H.H., 1997. Landslide risk—systematic approaches to assessment and management. Pp. 25-50 in D.M. Cruden and R.F. Fell (Eds.), *Landslide Assessment.* Rotterdam: A.A. Balkema.

Ellen, S.D., R.K. Mark, S.H. Cannon, and D.K. Knifong, 1993. *Map of Debris Flow Hazard in the Honolulu District of Oahu, Hawaii.* U.S. Geological Survey Open-File Report 93-213.

Ellen, S.D., R.K. Mark, G.F. Wieczorek, C.M. Wentworth, D.W. Ramsey, and T.E. May, 1997. *Map Showing Principal Debris-Flow Source Areas in the San Francisco Bay Region, California.* U.S. Geological Survey Open-File Report 97-745E, map scales 1:275,000 and 1:125,000.

FEMA (Federal Emergency Management Agency), 2003. *FEMA—Project Impact.* Online; available at http://www.app1.fema.gov/impact/; accessed June 2003.

Fleming, R.A., and F.A. Taylor, 1980. *Estimating the Costs of Landslide Damages in the United States,* U.S. Geological Survey Circular 832, 21 pp.

Geonet, 2003. *About landslides.* Online; available at http://www.geonet.org.nz/aboutlandslides/html; accessed June 2003.

Godschalk D.R., T. Beatley, P. Berke, D.J. Brower, and E.J. Kaiser, 1999. *Natural Hazard Mitigation, Recasting Disaster Policy and Planning.* Washington, D.C.: Island Press.

Godt, J.W., and W.Z. Savage, 1999. El Niño 1997-98: direct costs of damaging landslides in the San Francisco Bay Region. Pp. 47-55 in S. Griffiths and M.R. Stokes (Eds.), *Landslides. Proceedings 9th International Conference and Field Workshop on Landslides,* Bristol, U.K.

Gori, P.L., S.P. Jeer, and L.M. Highland, 2003. Enlisting the support of land-use planners to reduce debris-flow hazards in the United States. Pp. 1119-1127 in D. Rickenmann and C.-L. Chen (Eds.), *Debris-Flow Hazards Mitigation: Mechanics, Prediction, and Assessment. Proceedings of the Third International Conference on Debris-Flow Hazards Mitigation,* Davos, Switzerland, September 10-12, 2003. Rotterdam: Millpress.

Hansen, A., 1984. Landslide hazard analysis. Pp. 523-602 in D. Brunsden and D.B. Prior (Eds.), *Slope Instability,* New York: John Wiley & Sons.

Hansen, P.J., 1989. Grand Coulee Dam. *Engineering Geology in Washington, Vol. 1. Washington Division of Geology and Earth Resources Bulletin,* 78: 419-430.

Harp, E.L., and R.W. Jibson, 1995. *Inventory of Landslides Triggered by the 1994 Northridge, California Earthquake.* U.S. Geological Survey Open-File Report 95-213.

Harp, E.L., R.W. Jibson, W.Z. Savage, L.M. Highland, R.A. Larson, and S.S. Tan, 1999. Landslides triggered by January and March 1995 storms in southern California. Pp. 268-273 in K. Sassa (Ed.) *Landslides of the World.* Kyoto University Press.

Ho, K.K.S., E. Leroi, and W.J. Roberds, 2000. Quantitative risk assessment—applications, myths and future direction. Pp. 269-312 in *Proceedings International Conference on Geotechnical and Geological Engineering* (GeoEng 2000), Melbourne.

Iverson, R.M., 1997. The Physics of Debris Flows. *Reviews of Geophysics,* 35: 245-296.

Jackson, M.E., P.W. Bodin, W.Z. Savage, and E.M. Nel, 1996. Measurement of local horizontal velocities on the Slumgullion landslide using the global positioning system. Chapter 15 in D.J. Varnes and W.Z. Savage (Eds.), *The Slumgullion Earth Flow: A Large-Scale Natural Laboratory.* U.S. Geological Survey Bulletin 2130.

Jibson, R.W., 1992. The Mameyes, Puerto Rico, landslide disaster of October 7, 1985. In: J.A. Johnson and J.E. Slosson (Eds.), Landslides/Landslide Mitigation. *Reviews in Engineering Geology,* 9(5): 37-54.

Jibson, R.W., E.L. Harp, E. Schneider, R.A. Hajjeh, and R.A. Spiegel, 1998. An outbreak of coccidioidomycosis (valley fever) caused by landslides triggered by the 1994 Northridge, California earthquake. In: C.W. Welby and M.E. Gowan (Eds.), A Paradox of Power: Voices of Warning and Reason in the Geosciences. *Reviews in Engineering Geology,* 12: 53-61.

Jochim, C.L., W.P. Rogers, J.O. Truby, R.L. Wold, G. Weber, and S.P. Brown, 1988. *Colorado Landslide Hazard Mitigation Plan*. Colorado Geological Survey Bulletin 48, 149 pp.

Kockelman, W.J., 1975. *Use of USGS Earth Science Products by City Planning Agencies in the San Francisco Bay Region, California*. U.S. Geological Survey Open-File Report 75-276, 110 pp.

Larsen, M.C., and A.J. Torres Sanchez, 1992. Landslides Triggered by Hurricane Hugo in Eastern Puerto Rico, September 1989. *Caribbean Journal of Science*, 28: 113-125.

Liam Finn, W.D., 2003. Landslide-Generated Tsunamis: Geotechnical Considerations. *Pure and Applied Geophysics*, 160(10-11): 1879-1894.

Lo, D.O.K., 2001. Interim review of pilot applications of quantitative risk assessment to landslide problems in Hong Kong. Technical Note TN4/2001. Hong Kong Special Administrative Region.

LSU (Louisiana State University) CADGIS Research Laboratory, 2003. *Atlas: the Louisiana statewide GIS*. Online; available at http://atlas.lsu.edu /; accessed November 2003.

Lutton, R.J., D.C. Banks, and W.E. Strohm, Jr., 1979. Slides in Gaillard Cut, Panama Canal Zone. Pp. 151-224 *in* B. Voight (Ed.), *Rockslides and Avalanches*, Vol. 2. Elsevier.

Madej, M.A., 1995. Changes in channel-stored sediment, Redwood Creek, Northwestern California, 1947-1980. Chapter O *in* K.M. Nolan, H.M. Kelsey, and D.C. Marron (Eds.), *Geomorphic Processes and Aquatic Habitat in the Redwood Creek Basin, Northwestern California*. U.S. Geological Survey Professional Paper 1454.

Malone, A.W., 1998. Risk management and slope safety in Hong Kong. Pp. 3-17 *in* K.S. Li, J.N. Kay and K.K.S. Ho (Eds.), *Slope Engineering in Hong Kong*, Rotterdam: A.A. Balkema.

May, P.J., and R.E. Deyle, 1998. Governing land use in hazardous areas with a patchwork system. Pp. 57-82 *in* R.J. Burby (Ed.), *Cooperating with Nature, Confronting Natural Hazards with Land-Use Planning for Sustainable Communities*. Washington D.C.: Joseph Henry Press.

May, P.J., and W. Williams, 1986. *Disaster Policy Implementation, Managing Programs Under Shared Governance*. New York: Plenum Press.

McClelland, D.E., R.B. Foltz, C.M. Falter, C.D. Wilson, T. Cundy, R.L. Schuster, J. Saurbier, C. Rabe, and R. Heinemann, 1999. Relative Effects on a Low-Volume Road System of Landslides Resulting from Episodic Storms in Northern Idaho. *Transportation Research Record*, 1652: 235-243.

McKean, J., and J. Roering, 2003. Objective landslide detection and surface morphology mapping using high-resolution airborne laser altimetry. *Geomorphology*. Online; available at http://www.sciencedirect.com/; accessed December 2003.

Megahan, W.F., N.F. Day, and T.M. Bliss, 1978. Landslide occurrence in the western and central northern Rocky Mountain physiographic province in Idaho. Pp. 116-139 *in* C.T. Youngberg (Ed.), *Forest Soils and Land Use*. Proceedings, 5th North American Forestry Soils Conference, Colorado State University, Fort Collins, Colorado.

Morgan Hill, 1994. Ordinance No. 1176 N.S., An Ordinance of the City Council of the City of Morgan Hill Establishing Requirements and Standards for Geotechnical Review for Development Applications.

NCGS (North Carolina Geodetic Survey), 2003. *North Carolina floodplain mapping info*. Online; available at http://www.ncgs.state.nc.us/floodmap.html; accessed July 2003.

Newman, E.B., A.R. Paradis, and E.E. Brabb, 1978. Feasibility and Cost of Using a Computer to Prepare Landslide Susceptibility Maps of the San Francisco Bay Region, California. *U.S. Geological Survey Bulletin 1443*.

NHC (Natural Hazards Center), 2003. *Natural Hazards Center at the University of Colorado, Boulder*. Online; available at http://www.colorado.edu/hazards/; accessed June 2003.

NRC (National Research Council), 1983. *Risk Assessment in the Federal Government: Managing the Process*. Washington, D.C.: National Academy Press, 128 pp.

NRC (National Research Council), 1985. *Reducing Losses from Landslides in the United States.* Washington, D.C.: National Academy Press, 41 pp.

NRC (National Research Council), 1994. *Promoting the National Spatial Data Infrastructure Through Partnerships.* Washington, D.C.: National Academy Press, 128 pp.

NRC (National Research Council), 1996. *Understanding Risk, Informing Decisions in a Democratic Society.* Paul C. Stern and Harvey V. Fineberg (Eds.). Washington, D.C.: National Academy Press, 249 pp.

NRC (National Research Council), 1999. *The Impacts of Natural Disasters, A Framework for Loss Estimation.* Washington, D.C.: National Academy Press, 80 pp.

NRC (National Research Council), 2001. *National Spatial Data Infrastructure Partnership Programs, Rethinking the Focus.* Washington, D.C.: National Academy Press, 94 pp.

Olshansky, R.B., 1996. Financing Landslide Hazard Mitigation in the United States. *Journal of Environmental Planning and Management,* 39(3): 371-385.

OMB (Office of Management and Budget), 2002. *Circular A-16 revised.* Online; available at http://www.whitehouse.gov/omb/circulars/a016/a016_rev.html; accessed June 2003.

O'Tousa, J., 1995. La Conchita Landslide, Ventura County, California. *AEG News,* 38(4): 22-24.

Paterson R.G., 1998. The third sector: evolving partnerships in hazard mitigation. Pp. 203-230 *in* R.J. Burby (Ed.), *Cooperating with Nature, Confronting Natural Hazards with Land-Use Planning for Sustainable Communities.* Washington D.C.: Joseph Henry Press.

Pike, R.J., 1997. *Index to Detailed Maps of Landslides in the San Francisco Bay Region, California.* U.S. Geological Survey Open-File Report 97-745 D, map scale 1:275,000.

Radbruch-Hall, D.H., and C. Wentworth, 1971 *Estimated Relative Abundance of Landslides in the San Francisco Bay Region, California,* USGS Open-File Map, 1:500,000 scale.

Radbruch-Hall, D.H., R.B. Colton, W.E. Davies, I. Lucchitta, B.A. Skipp, and D.J. Varnes, 1982. *Landslide overview map of the conterminous United States,* U.S. Geological Survey Professional Paper 1183, 25 pp. Online; available at http://pubs.usgs.gov/pp/1983/p1183/ pp1183.html; accessed June 2003.

RMRC (Risk Management Resource Center), 2003. *Risk Management Resource Center.* Online; available at http://www.eriskcenter.org/; accessed June 2003.

Roberds, W.J., K. Ho, and K.W. Leung, 1997. An integrated methodology for risk management for development below potential natural terrain landslides. Pp. 333-346 *in* D. Cruden and R. Fell (Eds.), *Landslide Risk Assessment.* Rotterdam: A.A. Balkema.

Robinson, C.S., and D.B. Cochran, 1983. Engineering geology of Vail Pass I-70. Pp. 116-135 *in* J.L. Hynes (Ed.), *Engineering Geology and Environmental Constraints—Proceedings of the 33rd Annual Highway Geology Symposium.* Colorado Geological Survey Special Publication 22.

Rogers, W.P., 2003. *Critical Landslides of Colorado—A Year 2002 Review and Priority List.* Colorado Geological Survey Open-File Report OF-02-16, 1 map.

San Francisco Chronicle, 1983. Highway 50 reopens and Tahoe rejoices. June 24, p. 2.

Schmidt, K. M., J.J. Roering, J.D. Stock, W.E. Dietrich, D.R. Montgomery, and T. Schaub, 2001. Root cohesion variability and shallow landslide susceptibility in the Oregon Coast Range. *Canadian Geotechnical Journal,* 38: 995-1024.

Schuster, R.L., 1996. Chapter 2: Socioeconomic significance of landslides. Pp. 12-35 *in* A.K. Turner and R.L. Schuster (Eds.), *Landslides: Investigation and Mitigation.* Special Report 247, Transportation Research Board, National Research Council, Washington, D.C.: National Academy Press.

Schuster, R.L., 2001. Landslides: Effects on the Natural Environment. *Proceedings, Symposium on Engineering Geology and the Environment, International Association of Engineering Geologists,* 5: 3371-3387.

Schuster, R.L., and R.W. Fleming. 1986. Economic Losses and Fatalities from Landslides. *Bulletin Association of Engineering Geologists,* 23: 11-28.

Schuster, R.L., and L.M. Highland, 2001. *Socioeconomic and Environmental Impacts of Landslides in the Western Hemisphere*. U.S. Geological Survey Open-File Report 01-0276.

Scott, G.R., 1972. *Map Showing Landslides and Areas Susceptible to Landslides in the Morrison Quadrangle, Jefferson County, Colorado*. U. S. Geological Survey Miscellaneous Geologic Investigations Map I-790-B, scale 1", 24,000.

Sidle, R. C., A.J., Pearce, and C.L. O'Loughlin, 1984. *Hillslope Stability and Land Use*. Water Resources Monograph Series 11. Washington D.C.: American Geophysical Union, 140 pp.

Slosson, J.E., and J.P. Krohn. 1982. Southern California landslides of 1978 and 1980. Pp. 291-319 *in Proceedings, Floods and Debris Flows in Southern California and Arizona*. National Research Council, Environmental Quality Laboratory, California Institute of Technology, Pasadena.

Smith, T.C., 1982. Lawsuits and Claims Against Cities and Counties Mount After January 1982 Storm. *California Geology*, 35(7): 163-164.

Soeters, R., and C.J. van Westen, 1996. Chapter 8: Slope instability recognition, analysis, and zonation. Pp. 129-177 *in* A.K. Turner and R.L. Schuster (Eds.), *Landslides: Investigation and Mitigation*. Special Report 247, Transportation Research Board, National Research Council, Washington, D.C.: National Academy Press.

Solid Earth Science Working Group, 2002. *Living on a Restless Planet*. National Aeronautics and Space Administration and Jet Propulsion Laboratory, California Institute of Technology, 63 pp.

Spangle Associates, 1988. Geology and Planning: The Portola Valley Experience. William Spangle and Associates, Inc.

Spiker, E.C., and P.L. Gori, 2000. *National Landslide Hazards Mitigation Strategy—A Framework for Loss Reduction*, U.S. Geological Survey Open-File Report 00-450, 49 pp.

Spiker, E.C., and P.L. Gori, 2003. *National Landslide Hazards Mitigation Strategy—A Framework for Loss Reduction*, U.S. Geological Survey Circular 1244, 56 pp.

Stevenson, P.C., 1977. An Empirical Method for the Evaluation of Relative Landslide Risk. *Bulletin International Association of Engineering Geology*, 16: 69-72.

Swanston, D.N., 1991. Chapter 5, Natural Processes. Pp. 139-179 *in* W.R. Meehan (Ed.), *Influences of Forest and Rangeland Management on Salmonid Fishes and their Habitats*. American Fisheries Society Special Publication 19. Bethesda.

Swanston, D.N., and F.J. Swanson, 1976. Timber harvesting, mass erosion, and steepland forest geomorphology in the Pacific Northwest. Pp. 199-221 *in* D. R. Coates (Ed.), *Geomorphology and Engineering*. Stroudsburg: Dowden, Hutchinson & Ross.

Taylor, F.A., and E.E. Brabb, 1972. *Maps Showing Distribution and Cost by Counties of Structurally Damaging Landslides in the San Francisco Bay Region, California, Winter of 1968-69*. U.S. Geological Survey Miscellaneous Field Studies Map MF-327, scales 1:500,000, 1:1,000,000.

Turner, A.K., and R.L. Schuster (Eds.), 1996. *Landslides: Investigation and Mitigation*. Special Report 247, Transportation Research Board, National Research Council, Washington, DC, National Academy Press, 673 pp.

UN (United Nations), 2002. Living with Risk. A global review of disaster reduction initiatives. United Nations Inter-Agency Secretariat for the International Strategy for Disaster Reduction. preliminary version, Geneva, Switzerland, 384 pp.

University of California Committee on Cumulative Watershed Effects, 2001. *A scientific basis for the prediction of cumulative watershed effects*. Wildland Resources Center, Div. of Agriculture and Natural Resources, University of California, Berkeley. Report No. 46, 103 pp.

University of Utah, 1984. *Flooding and Landslides in Utah—An Economic Impact Analysis.* University of Utah Bureau of Economic and Business Research, Utah Department of Community and Economic Development, and Utah Office of Planning and Budget, Salt Lake City, 123 pp.

USACE (U.S. Army Corps of Engineers), 1983. *Embankment Criteria and Performance Report, Oahe Dam—Lake Oahe.* Department of the Army, U.S. Army Corps of Engineers, Omaha District.

USACE (U.S. Army Corps of Engineers), 1998. *Powerhouse Slope Landslide Position Paper, Oahe Dam—Lake Oahe, Missouri River, South Dakota.* Department of the Army, U.S. Army Corps of Engineers, Omaha District.

U.S. Bureau of Reclamation, 1936. *Grand Coulee Dam Project History,* Vol. IV. U.S. Bureau of Reclamation, Grand Coulee Project Office, Grand Coulee, Washington, 360 pp.

USGS (U.S. Geological Survey), 1974. San Francisco Bay Region Environment and Resources Planning Study—Plan for Completion of Study and Program Design for Fiscal Years 1974-1975. Unpublished, Menlo Park, California, 22 pp.

USGS (U.S. Geological Survey), 1997, *Introduction to the San Francisco Bay Region, California, Landslide Folio.* U.S. Geological Survey Open-File Report 97-745 A, 16 pp.

USGS (U.S. Geological Survey), 1999. *Real-Time Monitoring of Active Landslides.* U.S. Geological Survey Fact Sheet 091-99, 2 pp.

USGS (U.S. Geological Survey), 2003a. *Coastal erosion along the U.S. West Coast during the 1997-98 El Niño: expectations and observations.* Online; available at http://coastal.er.usgs.gov/lidar/ AGU fall98/; accessed July 2003.

USGS (U.S. Geological Survey), 2003b. *"Real-time" landslide monitoring—landslide instrumentation.* Online; available at http://vulcan.wr.usgs.gov/Projects/CalifLandslide/Maps/landslide_monitor. html; accessed July 2003.

USGS (U.S. Geological Survey), 2003c. *Landslide monitoring, Woodway, Washington.* Online; available at http://landslides.usgs.gov/woodway/; accessed June 2003.

USGS-HUD (U.S. Geological Survey and Department of Housing and Urban Development), 1971. *Program Design 1971, San Francisco Bay Region Environment and Resources Planning Study.* Menlo Park, California, 123 pp.

van Westen, C.J., 1993. *Application of Geographic Information Systems to Landslide Hazard Zonation.* ITC Publication No.15 (reprint of Ph.D. thesis). International Institute for Aerospace Survey and Earth Sciences (ITC), Enschede, The Netherlands, 245 pp.

Varnes, D.J., 1984. *Landslide Hazard Zonation: A Review of Principles and Practice.* Paris: UNESCO Press, 63 pp.

Walkinshaw, J., 1992. Landslide Correction Costs on U.S. State Highway Systems. *Transportation Research Record,* 1343: 36-41.

Wentworth, C.M., S.E. Graham, R.J. Pike, G.S. Beukelman, D.W. Ramsey, and A.D. Barron, 1997. *Summary Distribution of Slides and Earth Flows in the San Francisco Bay Region, California.* U.S. Geological Survey Open-File Report 97-745 C, map scales 1:275,000 and 1:125,000.

Wieczorek, G.F., 1982. *Map Showing Recently Active and Dormant Landslides near La Honda, Central Santa Cruz Mountains, California.* U.S. Geological Survey Miscellaneous Filed Studies Map MF-1422.

Wieczorek, G.F., 1984. Preparing a Detailed Landslide Inventory Map for Hazard Evaluation and Reduction. *Bulletin, Association of Engineering Geologists,* 21(3): 337-342.

Wieczorek, G.F., H.C. McWreath, and C. Davenport, 2001. *Remote Rainfall Sensing for Landslide Hazard Analysis.* U.S. Geological Survey Open-File Report 01-339.

Wieczorek, G.F., J.A. Coe, and J.W. Godt, 2003. Remote sensing of rainfall for debris-flow hazard analysis. *In* D. Rickenmann and C.-L. Chen (Eds.), *Debris-Flow Hazards Mitigation: Mechanics, Prediction, and Assessment. Proceedings of the Third International Conference on Debris-Flow Hazards Mitigation,* Davos, Switzerland, September 10-12. Rotterdam: Millpress.

Williams, G.P., and H.P. Guy, 1973. *Erosional and Depositional Aspects of Hurricane Camille in Virginia, 1969.* U.S. Geological Survey Professional Paper 804, 80 pp.

Wilson, D., R. Patten, and W.P. Megahan, 1982. Systematic Watershed Analysis Procedure for Clearwater National Forest. *Transportation Research Record,* 892: 50-56.

Wold, R.L.,Jr., and C.L. Jochim, 1989. *Landslide Loss Reduction: A Guide for State and Local Government.* Colorado Geological Survey Special Publication SP-33, 50 pp.

Works Bureau, 1998. *Information paper on slope safety, Provisional Legco Panel on Planning, Lands & Works. Hong Kong Legislative Council.* Online; available at http://www.legco.gov.hk/yr98-99/english/panels/plw/papers/pl23075b.htm; accessed October 2003.

Wright, S.G., and E.M. Rathje, 2003. Triggering Mechanisms of Slope Instability and Their Relationship to Earthquakes and Tsunamis. *Pure and Applied Geophysics,* 160(10-11): 1865-1877.

Wright, R.H., R.H. Campbell, and T.H. Nilsen, 1974. Preparation and use of isopleth maps of landslide deposits. *Geology,* 2: 483-485.

Wu, T.H., W.H. Tang, and H.H. Einstein, 1996. Chapter 6: Landslide Hazard and Risk Assessment. Pp. 106-118 *in* A.K. Turner and R.L. Schuster (Eds.), *Landslides: Investigation and Mitigation.* Special Report 247, Transportation Research Board, National Research Council, Washington, D.C.: National Academy Press.

Yim, K.P., S.T. Lau, and J.B. Massey, 1999. Community Preparedness and Response in Landslide Risk Reduction. *Proceedings of the Seminar on Geotechnical Risk Management in Hong Kong Institution of Engineers*: 145-155.

Zeizel, A.J., 1988. Foreword. Pp iii *in Colorado Landslide Hazard Mitigation Plan.* Colorado Geological Survey Bulletin 48.

Appendixes

APPENDIX A

Case Studies—A Widespread Problem

The following examples are presented to give the reader an awareness of the characteristics and variety of landslides—how they are triggered, their size and speed—and various community, institutional, or technological responses to these hazards.

Coastal Erosion, California. Much of the coastline of California consists of bluffs composed of relatively soft and poorly consolidated sediments (Figure A.1) that do not readily withstand erosion and undercutting by ocean waves and currents. The crest of the bluffs is an old, relatively level marine erosion surface that constitutes a highly desirable residential location, with easy access from inland and attractive ocean views. Many residents do not recognize the hazard of landslides and bluff retreat. Extensive damage results when winter storms and heavy rainfall combine to cause the coastal bluffs to fail. Recently, a small landslide on a coastal bluff received local notice; the following report was printed in the *North County Times* (San Diego County, California) on Saturday, May 25, 2002:

> DEL MAR—*Another chunk of the city's bluffs crashed onto the beach Friday morning, bringing the unstable sandstone much closer to the railroad tracks and creating a sheer precipice along a popular beach access point. The 20-foot-long, 5-foot-deep section slid from beneath a foot path at 11th Street at about 6:30 a.m. A pile of sandstone and sandstone boulders spread across the narrow beach below, nearly to the water's edge. Atop the 60-foot bluff, a number of deep fissures near the collapse suggest that more of the cliff may slide. The access trail at 11th Street is a favorite for many surfers and residents, although the city does not maintain the path, nor does it authorize its use.*

117

FIGURE A.1 Overview of the sea cliffs near Del Mar, California. The upper two-thirds of the cliff are formed of relatively soft, young (Pleistocene) Bay Point Formation, while the lower steep cliff is composed of more durable, older (Eocene) Torrey Sandstone. Both units retreat under wave attack and require costly retaining walls to provide some erosion protection.
SOURCE: Photo by Chris Metzler, Mira Costa College (http://www. miracosta.cc.ca.us/home/cmetzler/field_trip/top.html).

This landslide was modest in size; it injured not a soul and caused no property damage. Yet it occurred in a well-populated area and was witnessed by early morning beachgoers. The landslide hazard at the Del Mar bluffs persists, and the risk to the nearby railroad has increased.

Fatalities in Oregon. On November 18, 1996, heavy rains on the Oregon Coast Range dropped more than 18 cm of rain in a 24-hour period, leading to widespread shallow landsliding and debris flow generation. Along Rock Creek, near the town of Roseburg, people had built homes on the sloping surfaces of debris fans formed at the base of steep rocky hillslopes. When they built their houses, these hillslopes were covered with dense old forest. However, in the 1980s the slopes were clear-cut, despite concerns raised by Rock Creek residents, and waste wood was tossed into downslope steep gullies (Squier and Harvey, 2000). Removal of the trees caused the elaborate root systems that were laced through the stony, loose soil to fade away, as first the fine root hairs and then the larger roots died. With this loss of root strength, the soils became highly vulnerable to slope instability. In the late afternoon of November 18, the heavy rains of the day had progressively increased the soil water content until patches of soil in two small valleys failed and flowed downslope. The first failure tumbled down a canyon toward a house, but slowed and came to rest in wooded areas of the broad lower valley. The second tumbled down a steep canyon, where it picked up debris as it traveled at 5 to 7 m/s. By the time the debris flow had reached the low, gentler slopes on which houses were built, the mass had increased by 40 times. More than 4,000 m^3 of material, traveling upward of 9 m/s, smashed through one house, instantly killing the parents of a child who managed to run out the front door and escape as it swept by. The flow tore down the valley, sweeping away two more people and headed straight for another house. Fortunately it banked and turned to travel downstream, eventually coming to rest as it entered the mainstem Hubbard Creek. This entire event, from initial landslide, to the crushing death of four individuals, to the halting of the debris flow at Hubbard Creek, took only a few minutes. Subsequently, material disturbed by the slide, including wood and many household items (e.g., family pictures and clothing), continued down Hubbard Creek, depositing all the way out to the confluence with the Umpqua River (12 km from the headscarp) and beyond. This tragedy, and others in the storms of 1996 and 1997, led to the development of a warning system, compilation of hazard maps, and new legislation regarding forest practices.[1]

Rapid Debris Flow, New York. At approximately midday on April 27, 1993, a large landslide occurred along the foot of Bare Mountain in LaFayette, Onondaga County, New York, about 12 miles south of Syracuse. The landslide flowed rapidly toward the middle of the Tully Valley, involving approximately 50 acres of land. It destroyed three homes and caused the evacuation of four others. Luckily, most residents were away from their homes at the time, so no fatalities or serious injuries resulted from the landslide (Wieczorek et al., 1998b). The New York State Geological Survey reported that this was the largest landslide to have occurred in the state in more than 75 years. However, several parts of New York State, including the Finger Lakes region and the Hudson and Mohawk valleys, and other limited areas in the northeastern United States, such as Boston, and in the Puget Sound region of the Pacific Northwest are covered with similar glacial clays, originally deposited in lakes or in marine environments. Such clays are characterized by unstable internal structures and

---

[1]http://www.oregongeology.com/Landslide/Landslidehome.htm.

are prone to sudden landslide failures and rapid flowage of their debris, even over very gentle or level terrain. As a consequence, this type of landslide hazard became the subject of collaborative studies by the U.S. Geological Survey (USGS) and the State of New York (Wieczorek et al., 1996a). Onondaga County has begun to identify areas of landslide hazard and to zone them accordingly (Jäger and Wieczorek, 1994)

Huge Coastal Slide, Michigan. When local resident George Weeks walked his dog along the shore in Sleeping Bear Dunes National Lakeshore, Michigan, on an unusually warm February morning in 1995, he was shocked to find that where there had only recently been a beautiful beach was now a steep 100-foot drop into Lake Michigan (USGS, 1998; Figure A.2). The millions of cubic feet of sand that made up the beach and part of the high bluff above it had disappeared beneath the waters of the lake in a huge coastal landslide. Luckily, no one was on this popular beach when it slid off into the lake. The USGS was asked to investigate and determine the causes of the landslide. In 1997, USGS scientists studied the underwater part of the 1995 slide. Near the shore they found a deep hole where formerly there had been a gently sloping lake bottom. They also found that a thick blanket of slide debris extended more than 2 miles offshore

FIGURE A.2 Warning sign put up by National Park Service rangers immediately after the 1995 slide.
SOURCE: USGS (1998).

into water depths greater than 250 feet, much further and deeper than expected. Underwater video of the deeper part of the slide showed trees, which had been growing on the bluff and had been swept into deep water by the slide, protruding from the sand. The most probable cause was increased pore pressures within the bluff, resulting from entrapment of water from snowmelt behind the frozen bluff face or within confined sand layers. This mechanism is supported by history. Two similar landslides at Sleeping Bear Point—in December 1914 and March 1971—also occurred in unseasonably warm weather during winter months.

Sequential Coastal Slides, Massachusetts. Landslides are common along the Atlantic and Pacific coasts. An example from coastal southern California has been presented already. Cape Cod is formed of unconsolidated glacial deposits, and the onslaught of Atlantic waves and surges has caused the coastal bluffs on its outer arm to retreat in a series of landslides. The Highland Light lighthouse was built in 1797 and replaced in 1857, set back from the 183-foot bluff a distance of 500 feet. By 1990, a sequence of landslides caused the bluff to retreat to within 100 feet of the lighthouse. In 1996, the lighthouse was moved inland a distance of 450 feet. This sequence of events is an example of mitigation by "strategic retreat."

Madison County Debris Flows, Virginia. The foothills of the Blue Ridge in central Virginia are dotted with working farms along meandering rivers. The setting is peaceful to the casual observer, but closer inspection reveals massive boulders at the base of hills. These are signs of past violent geologic events—catastrophic large landslides and debris flows that have sculpted the local landscape. An intense storm on June 27th, 1995, produced 30 inches of rain in 16 hours over sections of the foothills of the Blue Ridge in Madison County, Virginia. Hundreds of debris flows were triggered on steep slopes and moved rapidly down mountain channels (Wieczorek et al., 1995). Small flows joined to form larger flows that, upon entering lowland valleys, spread mud, boulders, and other debris and inundated homes and farms (Figure A.3). One debris flow traveled nearly 2 miles, and an eyewitness estimated that it moved at a speed approaching 20 miles per hour. Because of the severity of the storm's effects, rural communities were isolated when bridges, roads, and power and telephone lines failed (Burton, 1996). The full extent of the damage was not recognized until aerial surveys were made several days later, and the county was declared a federal disaster area. Scientists have documented 51 historical debris flow events between 1844 and 1985 in parts of the Appalachians—most of them in the Blue Ridge area. Radiocarbon dating of plant remains from debris flow deposits indicate that these processes have occurred repeatedly over the last 34,000 years (USGS, 1996) and that

FIGURE A.3 Aerial view of Madison County, Virginia, debris flows; note destroyed house in upper right.
SOURCE: Morgan et al. (1999).

recurrence intervals for individual river basins are not more than 2,000 to 4,000 years (Eaton et al., 2003).

Multiple Landslide Types, Pacific Northwest. The Pacific Northwest coastal mountains are frequently subjected to episodes of numerous relatively small landslides following winter storms. A typical episode occurred after two separate regional storms in November and December 1998. The November storm triggered several small landslides in south King County, Washington, that blocked a few roads, including north-bound Interstate 5 near the Seattle airport, but caused no serious damage.

FIGURE A.4 Damage from the head scarp of a small earth slide closed Oregon State Highway 229, 13 miles north of Siletz, Lincoln County, Oregon. The landslide is typical of the numerous landslides caused by heavy winter rains in November and December 1998 in the Pacific Northwest.
SOURCE: Baum and Chleborad (1999).

The December storm was more serious. It followed a cold spell with considerable snow accumulation, and the runoff from snowmelt combined with rainfall caused flooding and triggered landslides throughout western Washington and, especially, western Oregon. A few of the landslides caused significant damage, and many temporarily blocked roads and highways (Figure A.4). USGS scientists conducted reconnaissance surveys to assess the landslide triggering mechanisms (Baum and Chleborad, 1999), and reported that there were a variety of landslide types, including earth slides, rock slides, rock falls, rapid earth flows, and debris flows.

Yosemite Valley Rock Falls, California. Rock falls and other types of mass movements are an important element of landscape development in Yosemite Valley. More than 400 rock falls have occurred in historic times; nine people have been killed and many others injured. The largest rock fall in memory occurred in July 1996, when two large rock blocks, with a combined weight of nearly 70,000 tons, fell more than 2,000 feet from the cliff face at Glacier Point to the valley floor near Happy Isles, a popular

FIGURE A.5 Rock fall damage to El Portal Road, Yosemite National Park, February 12, 2001.
SOURCE: National Park Service photo.

trailhead and concession stand (Wieczorek et al., 1992). The rock fall created an air blast that flattened about 2,000 trees in the vicinity. One person was killed at the concession stand, and 14 people were seriously injured. The National Park Service and USGS geologists conducted assessments and ultimately developed maps showing hazard zones and areas of rock fall potential in Yosemite Valley (Wieczorek et al., 1998a; Wieczorek and Snyder, 1999; Wieczorek et al., 1999). However these maps do not predict when or how frequently a rock fall will occur; consequently, neither the probability of a rock fall at any specific location nor the specific risk to people or facilities can be assessed.

Rock fall hazard is a continuing problem at Yosemite. On February 12, 2001, a rock fall closed the El Portal Road (Highway 140) approximately one-half mile east of the park boundary for about 24 hours. A slab of granite of unknown size was released approximately 1,000 feet above the road. On impact, the slab broke into many smaller pieces ranging in size from 2 to 12 feet in diameter, causing damage to the roadway (Figure A.5).

Earthquake-Induced Landslides. Earthquakes trigger landslides (NRC, 2003). There are many examples from Alaska, California, Montana, and other states prone to earthquakes. As well as the massive 1964 "Good Friday" Alaska earthquake, the 1994 magnitude 6.7 Northridge earthquake in southern California triggered more than 11,000 landslides—the vast majority were highly disrupted, shallow falls and slides of rock and debris that occurred over a wide area (Harp and Jibson, 1995). Landslide damage from the Northridge earthquake was only moderate because the area was neither heavily developed nor populated. However landslides did block roads, damage and destroy homes, disrupt transportation links and lifelines, and damage oil and gas production facilities.

Volcanic and Submarine Landslides. The slopes of volcanoes frequently experience landslides (e.g., Zimbelman et al., 2003). The cone of a volcano is built of material that falls or flows to the angle of repose, so that moderate shaking or subsequent eruptions can cause the volcano flanks to slide. The landslide hazard for volcanoes in the United States is geographically restricted and is confined to Alaska, California, Hawaii, Oregon, and Washington. In some cases, volcanic landslides can occur underwater, with the potential to cause destructive tsunamis (NRC, 2000). Giant slides have been identified surrounding the Hawaiian Islands (e.g., Moore and Clague, 2002) and on many other continental margins (e.g., Lewis and Collot, 2001). These landslides are among the largest known on earth, and most have occurred within the past 4 million years. Understanding giant submarine landslides is critically important because although they occur infrequently, they can produce destructive tsunamis and accordingly have the potential to cause enormous loss of life, property, and resources throughout surrounding coastal regions (e.g., the 1998 Papua New Guinea tsunami; Bardet et al., 2003; Dengler and Preuss, 2003; Liam Finn, 2003; Okal, 2003; Wright and Rathje, 2003).

REFERENCES

Bardet, J.P., C.E. Synolakis, H.L. Davies, F. Imamura, and E.A. Okal, 2003. Landslide Tsunamis: Recent Findings and Research Directions. *Pure and Applied Geophysics,* 160(10-11): 1793-1809.

Baum, R.L., and A.F. Chleborad, 1999. *Landslides triggered by Pacific Northwest Storms, November and December 1998.* U.S. Geological Survey web page, online; available at http://landslides.usgs.gov/Wash-Or/PNW98.html; accessed June 2003.

Burton, W.C., 1996. When the earth moved in Madison County. *Washington Post,* June 12.

Dengler, L., and J. Preuss, 2003. Mitigation Lessons from the July 17, 1998 Papua New Guinea Tsunami. *Pure and Applied Geophysics,* 160(10-11): 2001-2031.

Eaton, L.S., B.A. Morgan, R.C. Kochel, and A.D. Howard, 2003. Role of Debris Flows in Long-Term Landscape Denudation in the Central Appalachians of Virginia. *Geology,* 31: 339-342.

Harp, E.L., and R.W. Jibson, 1995. *Inventory of Landslides Triggered by the 1994 Northridge, California Earthquake.* U. S. Geological Survey Open-File Report 95-213.

Jäger, S., and G.F. Wieczorek, 1994. *Landslide Susceptibility in the Tully Valley Area, Finger Lakes Region, New York*. U.S. Geological Survey Open-File Report 94-615, 1 plate, scale 1:50,000.

Lewis, K., and J.-Y. Collot, 2001. Giant submarine avalanche: was this "deep impact" New Zealand style? *Water and Atmosphere. Online*, 9(1); available at http://www.niwa.co.nz/pubs/wa/091/avalanche. htm; accessed June 2003.

Liam Finn, W.D., 2003. Landslide-Generated Tsunamis: Geotechnical Considerations. *Pure and Applied Geophysics*, 160(10-11): 1879-1894.

Moore, J.G., and D.A. Clague, 2002. Mapping the Nuuanu and Wailau landslides in Hawaii. Pp. 223-244 *in* E. Takahashi, P.W. Lipman, M.O. Garcia, J. Naka, and S. Aramaki (Eds.), *Hawaiian Volcanoes: Deep Underwater Perspectives*. Geophysical Monograph 128, American Geophysical Union.

Morgan, B.A., G. Iovine, P. Chirico, and G.F. Wieczorek, 1999. *Inventory of Debris Flows and Floods in the Lovingston and Horseshoe Mountain, Va, 7.5' Quadrangles, from the August 19/20, 1969, Storm in Nelson County, Virginia*. U.S. Geological Survey Open-File Report 99-518.

NRC (National Research Council), 2000. *Review of the U.S. Geological Survey's Volcano Hazard Program*. Washington, D.C.: National Academy Press, 138 pp.

NRC (National Research Council), 2003. *Living on an Active Earth: Perspectives on Earthquake Science*. Washington, D.C.: National Academy Press, 365 pp.

Okal, E.A., 2003. T Waves from the 1998 Papua New Guinea Earthquake and Its Aftershocks: Timing the Tsunamigenic Slump. *Pure and Applied Geophysics*, 160(10-11): 1843-1863.

Squier, L.R., and A.F. Harvey, 2000. Two debris flows in Coast Range, Oregon, USA: logging and public policy impacts. Pp. 127-138 *in* G.F. Wieczorek and N.D. Naeser (Eds.), *Debris Flow Hazards Mitigation: Mechanics, Prediction and Assessment*. Rotterdam, The Netherlands: Balkema.

USGS (U.S. Geological Survey), 1996. *Debris-Flow Hazards in the Blue Ridge of Virginia*, U.S. Geological Survey Fact Sheet 159-96, 4 pp.

USGS (U.S. Geological Survey), 1998. *Popular Beach Disappears Underwater in Huge Coastal Landslide—Sleeping Bear Dunes, Michigan*. U.S. Geological Survey Fact Sheet 020-98, 2 pp.

Wieczorek, G.F., and J.B. Snyder, 1999. *Rock falls from Glacier Point Above Camp Curry, Yosemite National Park, California*. U.S. Geological Survey Open-File Report 99-385.

Wieczorek, G.F., J.B. Snyder, C.S. Alger, and K.A. Isaacson, 1992. *Rock Falls in Yosemite Valley, California*. U.S. Geological Survey Open-File Report 92-387, 38 pp.

Wieczorek, G.F., P.L. Gori, R.H. Campbell, and B.A. Morgan, 1995. *Landslide and Debris-Flow Hazards Caused by the June 27, 1995, Storm in Madison County, Virginia*. U.S. Geological Survey Open-File Report 95-822, 33 pp.

Wieczorek, G.F., P.L. Gori, S. Jäger, W.M. Kappel, and D. Negussey, 1996a. Assessment and management of landslide hazards near the Tully Valley Landslide, Syracuse, New York, USA. Pp. 411-416 *in Proceedings of the Seventh International Symposium on Landslides*, June 17-21, 1996, Trondheim, Norway.

Wieczorek, G.F., M.M. Morrissey, G. Iovine, and J. Godt, 1998a. *Rock-Fall Hazards in the Yosemite Valley*. U.S. Geological Survey Open-File Report 98-467.

Wieczorek, G.F., D. Negussey, and W.M. Kappel, 1998b. *Landslide Hazards in Glacial Lake Clays—Tully Valley, New York*. U.S. Geological Survey Fact Sheet 013-0098. Online; available at http://pubs.usgs.gov/fs/fs13-98/; accessed June 2003.

Wieczorek, G.F., M.M. Morrissey, G. Iovine, and J. Godt, 1999. *Rock-Fall Potential in the Yosemite Valley, California*. U.S. Geological Survey Open-File Report 99-578.

Wright, S.G., and E.M. Rathje, 2003. Triggering Mechanisms of Slope Instability and Their Relationship to Earthquakes and Tsunamis. *Pure and Applied Geophysics*, 160(10-11): 1865-1877.

Zimbelman, D., R.J. Watters, S. Bowman, and I. Firth, 2003. Quantifying Hazard and Risk Assessments at Active Volcanoes. *EOS*, 84(23): 213, 216-217.

Appendix B

Committee Biographies

J. Freeman Gilbert (NAS) is a research professor at the Scripps Institution of Oceanography, University of California, San Diego. His research interests include theoretical, inferential, and computational geophysics. He is one of the founders of the San Diego Supercomputer Center and the National Partnership for Advanced Computational Infrastructure, sponsored by the National Science Foundation.

William E. Dietrich (NAS) is professor of geomorphology at the University of California, Berkeley. He has appointments in the Earth and Planetary Science Department, the Department of Geography, and the Earth Sciences Division of Lawrence Berkeley National Laboratory. His current research includes mechanistic analysis of landscape processes and evolution, identifying linkages between ecological and geomorphic processes, as well as building tools to tackle pressing environmental problems.

J. Michael Duncan (NAE) is a University Distinguished Professor in the Department of Civil and Environmental Engineering at the Virginia Polytechnic Institute and State University. Dr. Duncan is a geotechnical engineer specializing in problems of soil-structure interaction, stability, and seepage.

Philip E. LaMoreaux (NAE) is now a hydrogeology and environmental geology consultant, after retiring from service as chief of the Groundwater Branch of the U.S. Geological Survey, as state geologist of Alabama, as professor of geology at the University of Alabama, and as director of the Environmental Institute for Waste Management Studies for Alabama.

George G. Mader is a city planner and president of Spangle Associates, Inc., a city planning and research consulting firm in the San Francisco Bay region. He has specialized in using city planning to reduce risks from geologic hazards. His activities have included teaching, research, and planning in this country and abroad.

William F. Marcuson III (NAE) is president of W.F. Marcuson III and Associates, Inc. and director emeritus of the Geotechnical Laboratory, U.S. Army Engineer Research and Development Center. His research activities have focused on experimental and analytical studies of soil behavior related to geotechnical engineering problems, seismic design, analysis, and remediation of embankment dams, and seismically induced liquefaction of soils.

Peter J. May is professor of political science at the University of Washington. His research is concerned with regulatory policy design and implementation, with particular attention to environmental regulation and policy making regarding natural hazards.

Norbert R. Morgenstern (NAE) is a University Professor of Civil Engineering (emeritus) at the University of Alberta and an internationally recognized authority in the field of geotechnical engineering. He has considerable experience with landslides at both theoretical and applied levels.

Jane Preuss is a principal with GeoEngineers, a company specializing in geotechnical engineering and engineering geology. She has more than 20 years of experience as a practicing urban planner, working with clients from both public and private sectors. Her main areas of interest include land-use and environmental planning for mitigation and preparedness against the effects of natural hazards such as floods, landslides, earthquakes, tsunamis, and high winds.

A. Keith Turner holds concurrent appointments as professor of geological engineering at the Colorado School of Mines and professor of engineering geology at Delft University of Technology in The Netherlands. His chief research interest involves computer applications to geological and environmental studies, including landslide assessments in Colorado and Canada.

T. Leslie Youd is professor of civil engineering at Brigham Young University, where he teaches courses in geotechnical and earthquake engineering and conducts research on liquefaction and ground failure. Dr. Youd was formerly (1967 to 1984) a research civil engineer with the U.S. Geological Survey, Menlo Park, California.

NATIONAL RESEARCH COUNCIL STAFF

David A. Feary is a senior staff scientist with the National Research Council's Board on Earth Sciences and Resources. His research activities have focused on the geological and geophysical evolution of continental margins, particularly the factors controlling carbonate deposition and reef development within different climatic regimes.

APPENDIX C

Acronyms

AASG	Association of American State Geologists
AEG	Association of Engineering Geologists
AGI	American Geological Institute
AGS	Australian Geomechanics Society
AIPG	American Institute of Professional Geologists
ALACE	Airborne LIDAR Assessment of Coastal Erosion
APA	American Planning Association
ASCE	American Society of Civil Engineers
ATM	airborne topographic mapper
BLM	Bureau of Land Management
CGS	Colorado Geological Survey
dGPS	differential global positioning system
DOGAMI	Oregon Department of Geology and Mineral Industries
DOI	Department of the Interior
EERI	Earthquake Engineering Research Institute
FEMA	Federal Emergency Management Agency
FGDC	Federal Geographic Data Committee
FHWA	Federal Highway Administration
FRA	Federal Railway Administration

GHAD	Geologic Hazard Abatement District
GIS	geographic information system
GPS	global positioning system
HMGP	Hazard Mitigation Grant Program
HUD	Department of Housing and Urban Development
IAEG	International Association for Engineering Geology and the Environment
ICL	International Consortium on Landslides
ICSSC	Interagency Committee of Seismic Safety in Construction
InSAR	interferometric synthetic aperture radar
IPA	Intergovernmental Personnel Act
IPL	International Programme on Landslides
ISDR	International Strategy for Disaster Reduction
ISRM	International Society for Rock Mechanics
ISSMGE	International Society for Soil Mechanics and Geotechnical Engineering
JTC-1	Joint Technical Committee on Landslides
LFL	Learning from Landslides
LIDAR	light detection and ranging
NASA	National Aeronautics and Space Administration
NCALM	NSF-sponsored Center for Airborne Laser Mapping
NEHRP	National Earthquake Hazard Reduction Program
NEXRAD	next-generation radar
NFIP	National Flood Insurance Program
NOAA	National Oceanic and Atmospheric Administration
NOS	National Ocean Service
NPS	National Park Service
NRC	National Research Council
NSDI	National Spatial Data Infrastructure
NSF	National Science Foundation
NWS	National Weather Service
OES	California Office of Emergency Services
UNESCO	United Nations Educational, Scientific and Cultural Organization
USACE	U.S.Army Corps of Engineers
USFS	U.S. Forest Service
USGS	U.S.Geological Survey